METHODS IN COMPUTATIONAL PHYSICS
Advances in Research and Applications

Volume 14

Radio Astronomy

Methods in Computational Physics

Advances in Research and Applications

1   STATISTICAL PHYSICS
2   QUANTUM MECHANICS
3   FUNDAMENTAL METHODS IN HYDRODYNAMICS
4   APPLICATIONS IN HYDRODYNAMICS
5   NUCLEAR PARTICLE KINEMATICS
6   NUCLEAR PHYSICS
7   ASTROPHYSICS
8   ENERGY BANDS OF SOLIDS
9   PLASMA PHYSICS
10   ATOMIC AND MOLECULAR SCATTERING
11*   SEISMOLOGY: SURFACE WAVES AND EARTH OSCILLATIONS
12*   SEISMOLOGY: BODY WAVES AND SOURCES
13*   GEOPHYSICS
14   RADIO ASTRONOMY

* Volume Editor: Bruce A. Bolt.

METHODS IN COMPUTATIONAL PHYSICS

Advances in Research and Applications

Editors

BERNI ALDER

*Lawrence Livermore Laboratory
Livermore, California*

SIDNEY FERNBACH

*Lawrence Livermore Laboratory
Livermore, California*

MANUEL ROTENBERG

*University of California
La Jolla, California*

Volume 14

Radio Astronomy

1975

ACADEMIC PRESS   NEW YORK   SAN FRANCISCO   LONDON

A Subsidiary of Harcourt Brace Jovanovich, Publishers

Copyright © 1975, by Academic Press, Inc.
ALL RIGHTS RESERVED.
NO PART OF THIS PUBLICATION MAY BE REPRODUCED OR
TRANSMITTED IN ANY FORM OR BY ANY MEANS, ELECTRONIC
OR MECHANICAL, INCLUDING PHOTOCOPY, RECORDING, OR ANY
INFORMATION STORAGE AND RETRIEVAL SYSTEM, WITHOUT
PERMISSION IN WRITING FROM THE PUBLISHER.

ACADEMIC PRESS, INC.
111 Fifth Avenue, New York, New York 10003

United Kingdom Edition published by
ACADEMIC PRESS, INC. (LONDON) LTD.
24/28 Oval Road, London NW1

LIBRARY OF CONGRESS CATALOG CARD NUMBER: 63-18406

ISBN 0-12-460814-0

PRINTED IN THE UNITED STATES OF AMERICA

Contents

CONTRIBUTORS vii
PREFACE ix

RADIOHELIOGRAPHY

N. R. Labrum, D. J. McLean, and J. P. Wild

I. Introduction	1
II. The Sun and Its Radio Image	2
III. The Principles of Radioheliography	4
IV. The Evolution of the Radioheliograph	13
V. The Culgoora Radioheliograph	51
VI. Culgoora Data Processing	27
VII. Solar Radio Astronomy with the Radioheliograph	44
VIII. Future Developments in Radioheliography	15
References	52

PULSAR SIGNAL PROCESSING

Timothy H. Hankins and Barney J. Rickett

I. Introduction	56
II. Pulsar Searches	57
III. Dispersion	64
IV. Sampling, Resolution, and Average Profiles	81
V. Polarization	84
VI. Intensity Variations with Time	87
VII. Intensity Variations with Frequency	104
VIII. Interstellar Scattering and Scintillation	109
IX. Timing Measurements	118
References	126

APERTURE SYNTHESIS

W. N. Brouw

I. Introduction	131
II. Aperture Synthesis	132
III. Earth Rotation Aperture Synthesis	139
IV. Data Processing	160
V. Conclusion	173
References	173

COMPUTATIONS IN RADIO-FREQUENCY SPECTROSCOPY

John A. Ball

I. Introduction to Spectroscopy in Radio Astronomy	177
II. Power Spectra	178
III. Selected Problems in Calibration and Observing Techniques	200
IV. Selected Problems in Interpretation of Spectra	208
References	218

AUTHOR INDEX	221
SUBJECT INDEX	225
CONTENTS OF PREVIOUS VOLUMES	232

Contributors

Numbers in parentheses indicate the pages on which the authors' contributions begin.

JOHN A. BALL, *Center for Astrophysics, Harvard College Observatory and Smithsonian Astrophysical Observatory, Cambridge, Massachusetts* (177)

W. N. BROUW, *Netherlands Foundation for Radio Astronomy and Leiden Observatory, Leiden, The Netherlands* (131)

TIMOTHY H. HANKINS,[*] *Department of Applied Physics and Information Science, University of California at San Diego, La Jolla, California* (55)

N. R. LABRUM,[†] *Division of Radiophysics, CSIRO, Sydney, Australia* (1)

D. J. MCLEAN,[†] *Division of Radiophysics, CSIRO, Sydney, Australia* (1)

BARNEY J. RICKETT, *Department of Applied Physics and Information Science, University of California at San Diego, La Jolla, California* (55)

J. P. WILD,[†] *Division of Radiophysics, CSIRO, Sydney, Australia* (1)

[*] Present address: Arecibo Observatory, Box 995, Arecibo, Puerto Rico 00612.
[†] Mailing adress: P.O. Box 76, Epping, N.S.W. 2121, Australia.

Preface

UP UNTIL THE EARLY 1930's, when K. G. Jansky discovered galactic radio noise, astronomers relied solely upon the "optical window" in the electromagnetic spectrum (400 to 800 nanometers) for their observations. Techniques developed since the Second World War have brought other parts of the spectrum into use. In particular, the "radio window" (roughly 0.01 to 10 meters) has yielded spectacular results in furthering our knowledge of stellar and interstellar physics.

The sharp change in wavelength gives astronomers access to new astrophysical phenomena and the fivefold increase in bandwidth offers a diversity of observations that is not available in optical astronomy. But this new freedom is not without its cost, because technologies developed for the optical window are not appropriate to the radio window, the larger wavelength diminishes the resolving power of the telescope, and the information density and signal integrating memory provided by photographic film are not available. These are old problems that need new solutions. But new problems have also arisen in the form of enormous data rates, vanishingly small signal-to-noise ratios, and the requirements for high-accuracy time resolution. The present volume is devoted to the solution of some of these problems, with emphasis on the role played by the digital computer both as a control device and as a calculator.

The demands of both spatial resolution and high data rates become evident in radioheliography, where fast moving phenomena of small size are of interest. The mathematical intricacies of the problem and the prototype radioheliograph at Culgoora are described in the article on radioheliography.

Pulsar signals pose the problem of time resolution and signal amplification. Peculiar to pulsars are their very stable periods, even though the rate varies greatly from pulsar to pulsar. The intensity of a given pulsar, on the other hand, varies enormously. The sampling of pulsar signals adequate to resolve time structure to the accuracy of microseconds presents data handling problems of major proportion. The paper on pulsars discusses the averaging methods used to cope with this problem.

As already mentioned, the long wavelength of the radio window does not permit high angular resolution without some kind of subterfuge, since the simple scaling up of antenna size is not practical. One of the techniques used to increase the effective size of the antenna is aperture synthesis; the mathematical basis and the computational problems of this method are the subject of a third article.

Finally, we include a paper on radio-frequency spectroscopy. The discussion bears on the problems of data acquisition, reduction and interpretation of spectral-line emission in the radio-frequency spectrum.

METHODS IN COMPUTATIONAL PHYSICS

Advances in Research and Applications

Volume 14

Radio Astronomy

Radioheliography

N. R. LABRUM, D. J. MCLEAN, AND J. P. WILD

DIVISION OF RADIOPHYSICS, CSIRO
SYDNEY, AUSTRALIA*

I. Introduction 1
II. The Sun and Its Radio Image 2
III. The Principles of Radioheliography 4
    A. General Requirements 4
    B. The Geometry of Spaced Arrays 6
    C. Processing the Received Signals 10
IV. The Evolution of the Radioheliograph 13
V. The Culgoora Radioheliograph 15
    A. General Description 15
    B. Array Beam Formation 16
    C. Picture Formation 20
    D. Array Beam Steering 21
    E. The Receiving System 22
    F. The Phase-Control System 23
VI. Culgoora Data Processing 27
    A. Off-Line Processing of Culgoora Data 28
    B. On-Line Data Handling 38
VII. Solar Radio Astronomy with the Radioheliograph . . . 44
    Radio Bursts from the Corona 45
VIII. Future Developments in Radioheliography 51
    References 52

I. Introduction

RADIOHELIOGRAPHY IS THE BRANCH of radio astronomy concerned with obtaining radio images of the Sun for the purpose of studying features of the Sun's atmosphere. We shall be less concerned in the present article with the contribution of radioheliography to solar physics (a brief review of selected results is included at the end of the paper) than with the special instrumental and computational problems which are posed. It is true in principle that obtaining a radio image of the Sun involves the same basic type of measurement as mapping any other part of the heavens at radio frequencies. However, the large and rapid variations of the solar emission create special problems for the solar radio astronomer which are not encountered in other branches of radio astronomy;

* *Mailing address:* P.O. Box 76, Epping, N.S.W. 2121, Australia.

this circumstance has led to the evolution of a distinctive breed of instrumentation and instrumental philosophy.

We shall begin this review with a brief description of the Sun and those of its characteristics which determine the nature of its image in the radio spectrum, and then discuss the general principles of radioheliography as they have evolved during the past 25 years. The most exacting type of radioheliograph is one which not only produces an image with high resolution, but also does this in a very short time so that "movies" can be made of fast-changing phenomena. At the time of writing only one such instrument has been put into service; this is the Culgoora radioheliograph in Australia, and a detailed description of this prototype instrument is given in Sections V and VI. We shall be concerned not only with a technical description of the hardware but also with the special data-handling problems which result from the flood of data supplied at a rate not normally encountered in observational astronomy. Solutions to these problems have been evolved as a result of operational experience. In Section VII we present a selection of observational results obtained with the Culgoora radioheliograph and touch briefly on the significance of these results to solar physics and to astrophysics in general.

II. The Sun and Its Radio Image

The distinctive requirements of a radioheliograph are a consequence of the physical nature of the Sun itself, and so we should first consider the main features of our object of study. The visible Sun, a ball of dense gas 1.4×10^6 km in diameter, subtends an angle of about $\frac{1}{2}°$ at the Earth. This gas, intensely heated by thermonuclear burning of hydrogen in the central core, is held together by gravity which balances the expansive force of the gas pressure. The temperature at the visible surface of the photosphere is some $6000°K$. Above the photosphere there extends a vast outer "atmosphere" of tenuous gas (the corona). Although sophisticated optical instrumentation has been developed to study the chromosphere (the bright region just above the photosphere) and inner corona, the outer corona can be observed optically from the surface of the Earth only at the time of a total eclipse. On the other hand, the solar atmosphere can be studied by its natural radio emission at all times, so that radio astronomy now probably provides the most powerful technique for exploring the outer atmosphere of the Sun.

Magnetohydrodynamic waves generated in a turbulent region below the visible surface of the Sun pass through the chromosphere and heat the corona to such high temperatures ($\sim 10^6 °K$) that the gas is completely ionized. The electron density of this ionized gas decreases monotonically with height, as shown in Fig. 1. Now in an ionized gas with electron density N_e the only radio waves that can propagate are those with frequency f greater than a cutoff frequency f_p given by

$$f_p = e(N_e/\pi m \epsilon_0)^{1/2}, \tag{1}$$

where e and m are the electronic charge and mass, and ϵ_0 the permittivity of free space. Numerically,

$$f_p(\text{MHz}) = 9 \times 10^{-3}[N_e(\text{cm}^{-3})]^{1/2}. \tag{2}$$

Lower-frequency waves are reflected when they encounter dense gas where $f_p \geq f$, in the same way that terrestrial radio signals are bounced off the ionosphere. This means that different frequencies received from the Sun come from different layers in the atmosphere (cf. Fig. 1), and that radio-frequency studies of the corona will be quite unhampered by the "glare" which frustrates optical observation. Reference to Fig. 2 shows how the angular diameter of the radio Sun changes with wavelength: at millimeter and centimeter wavelengths the diameter is little greater than that of the optical disk, while at meter and decameter wavelengths it is considerably greater, being double the optical diameter at a frequency of about 40 MHz.

Solar activity associated with sunspots and related phenomena mani-

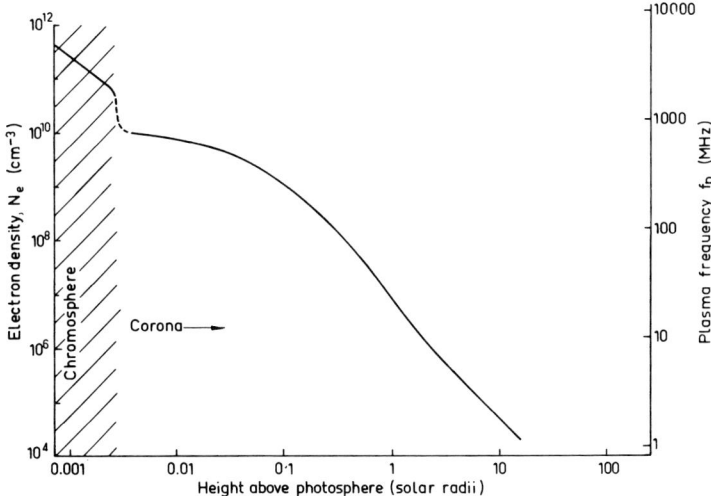

FIG. 1. Electron density and plasma frequency in the solar atmosphere. The coronal values are those deduced from radio observations and correspond to conditions within coronal streamers. The exact height of the sharp transition from corona to chromosphere is somewhat uncertain; we have used the values obtained by Chiuderi et al. (1971) for the transition in a solar active region.

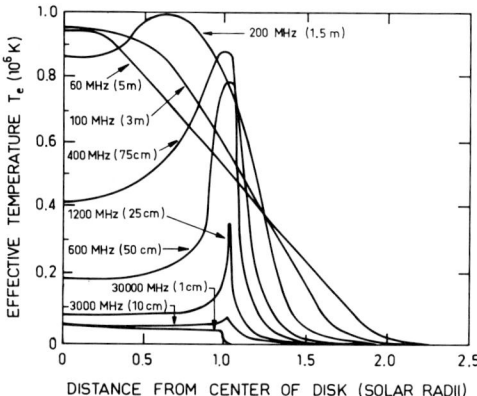

Fig. 2. The computed brightness distribution over the Sun for a series of radio frequencies. The values assumed for chromospheric and coronal temperatures are 3×10^4 and $10^6\,°K$, respectively (from Smerd, 1950).

fests itself in a number of ways in the radio spectrum, producing radiation which is highly variable, with time scales ranging from the very slowly varying phenomena associated with the 27-day rotation period of active centers to short intense bursts of radiation with time scales of minutes, seconds, or even fractions of a second. In extreme cases the intensity can vary by a factor of a thousand or so in as little as 1 sec. The most intense bursts of radiation are associated with solar flares. At meter wavelengths the intensity of emission from a localized region of the Sun may vary over a range of the order of $1:10^6$. This strong variability is caused by radiation generated by nonthermal processes, notably from plasma oscillations and synchrotron radiation. Magnetic fields play an important role in the generation and propagation of radio waves and certain components of the Sun's radiation show circular polarization in consequence. Important information is therefore available from a knowledge of the state of polarization of the radiation.

III. The Principles of Radioheliography

A. GENERAL REQUIREMENTS

The distinctive requirements of radioheliography which differ from those of general radio astronomy are threefold:

(a) Since the Sun is an intense isolated object in the sky the field to be imaged is confined to within specifiable angular bounds and all sources of radiation outside this field can be ignored.

(b) The wavelength of observation is determined by the height of the layer of the solar atmosphere to be studied. Therefore the observer can in no sense select his wavelength to minimize his instrumental problems. Indeed many of the most interesting phenomena to be studied occur at inconveniently long wavelengths.

(c) To register the rapidly varying solar phenomena a complete image must be built up in a very short time, typically 1 sec. The Sun must be tracked for long periods of time and the recording process must possess a dynamic range large enough to accommodate intensity variations of $10^6:1$.

The first of these requirements, (a), represents a relaxation of the general requirement for mapping in radio astronomy and introduces an elegant simplification into the design of aerial arrays. This was first recognized by Christiansen (1953) and by Tanaka and Kakinuma (1954); in building linear arrays to take one-dimensional strip scans of the radio Sun, these authors realized that the linear aperture need not be continuously filled but can consist of a series of discrete antennas with uniform spacing of as much as 60 wavelengths. Such an array is the radio analog of a diffraction grating, the antenna pattern consisting of a series of sharp well-spaced lobes (Fig. 3). If the angular spacing of the lobes is designed to exceed the angular diameter of the radio Sun no ambiguities arise; indeed repeated strip scans of the persistent features of the Sun's radiation can be obtained with extremely simple instrumentation by allowing the Sun to drift through one fringe after another and so recording a strip scan every 5 min or so.

The second requirement, (b), means that in order to study the important outer regions of the corona that are barely accessible to optical observation we need to observe at long wavelengths and so to use enormous antenna apertures. For instance, at a wavelength of 3 meters a modest angular resolution of 3′ arc (equivalent to one-tenth of the diameter of the optical disk) requires a 3-km aperture (since $3'$ arc $\approx 10^{-3}$ rad). This in turn means that on economic grounds we are led to the use of arrays of antenna elements—preferably "dilute" (as opposed to "filled") arrays. The first array designed specifically to record two-dimensional images of the Sun was built by Christiansen and Mathewson (1958) using a crossed grating of 64 parabolic antennas. This operated at the relatively short wavelength of 21 cm and so the overall dimensions were moderate (380 meters).

The third requirement, (c)—to register an almost instantaneous image (and so achieve in the radio spectrum what we take for granted in optical photographic instruments)—represents the biggest departure from the

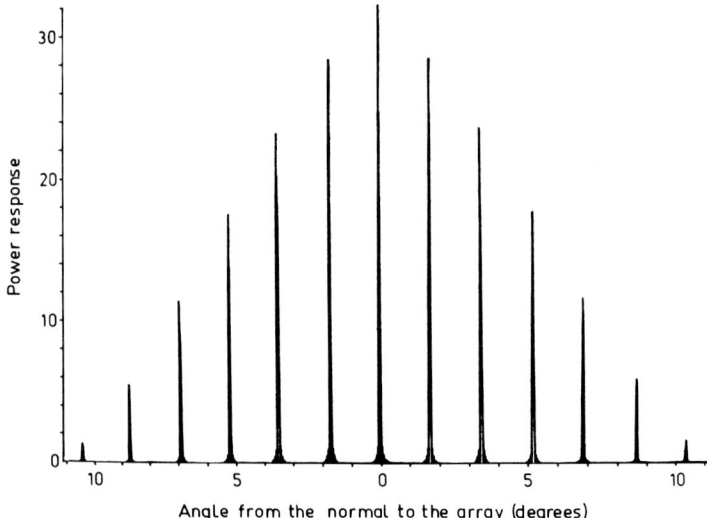

FIG. 3. The directional diagram of a 32-element grating interferometer, with spacing 33λ between adjacent elements. The outer envelope of the grating lobes represents the polar diagram of the individual aerial elements (Christiansen and Warburton, 1953).

normal high-resolution techniques of radio astronomy, and the greatest challenge. Conceptually the simplest method of fast image forming is to scan a pencil beam as in a television set. However, this technique is basically wasteful; to achieve highest sensitivity the information received at all picture points should ideally be gathered simultaneously rather than serially.

For a radioheliograph, then, the general requirements are (i) a frequency of operation suited to the region of the solar atmosphere to be studied; (ii) an array capable of mapping an isolated source of known extent; and (iii) a technique for processing the received signals with as many simultaneous outputs as practicable.

B. THE GEOMETRY OF SPACED ARRAYS

Consider first a uniform square aperture (Fig. 4a) with sides of length a wavelengths $(a \gg 1)$. To calculate the power polar diagram p we adopt the usual procedure of taking the two-dimensional Fourier transform* $\bar{u}$ of the aperture distribution u and squaring it:

$$p = (\bar{u})^2 \qquad (3)$$

* The Fourier transform $F(x, y)$ of the function $f(\xi, \eta)$ is defined by $F(x, y) = \iint f(\xi, \eta) \exp [2\pi i(x\xi + y\eta)] \, d\xi \, d\eta$.

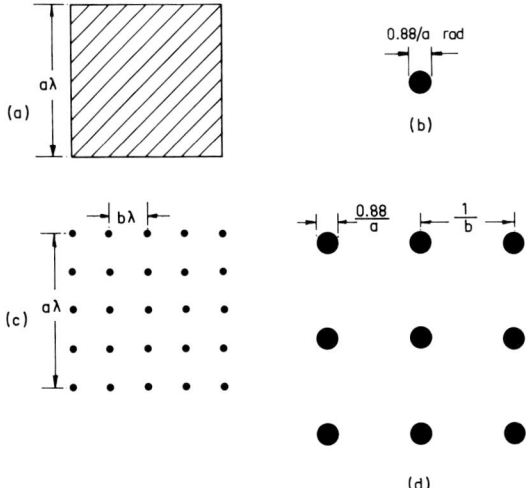

Fig. 4. (a) A uniformly illuminated square antenna aperture. (b) The power polar diagram corresponding to (a). This is a single pencil beam perpendicular to the plane of the aperture, with an angular width determined by the dimensions of the antenna. It is represented here by its half-power contour. (c) A uniformly spaced array of small (i.e., nondirectional) co-phased antennas, covering the same area as the filled aperture (a). (d) The central part of the power polar diagram of (c). This pattern is a matrix of pencil beams, each of the same form as the single beam in (b); the spacing between adjacent beams is inversely proportional to the distance between adjacent radiators in the array.

so obtaining a $[(\sin \pi ax)/\pi ax]^2 \cdot [(\sin \pi ay)/\pi ay]^2$ type of pattern, a cross-section of which is depicted in Fig. 4b. This is an example of a *continuous* filled-aperture array.

The simplest type of two-dimensional *spaced* array to consider is a uniform square matrix with sides of a wavelengths, made up of small antenna elements spaced b wavelengths apart $(a \gg b \gg 1)$ (Fig. 4c). The aperture distribution is the same as in the previous example except that it has been multiplied by a two-dimensional sampling function—a matrix of spikes with spacing b which covers an infinite field. The Fourier transform of the resulting product is obtained by applying the convolution theorem, which says that we must take the transform of the square aperture and *convolve* it with the Fourier transform of the sampling function. The latter is simply another spike function covering an infinite field with spacing $1/b$, and so the final result (the power response) is, as shown in Fig. 4d, a periodic pattern with period $1/b$ in which each element has the form $[(\sin \pi ax)/\pi ax]^2 \cdot [(\sin \pi ay)/\pi ay]^2$. Clearly when we use spaced arrays in radioheliography we should select b so that the angle $1/b$ rad is slightly greater than the largest solar image we expect to record. Typically we need fields of $\frac{1}{2}°$ to $2°$, depending on wavelength, and so the spacing between antenna elements should be around 30 to 100 wavelengths.

We now turn to another class of arrays in which a large part of the redundancy of the regular matrix is removed and far fewer elements are used in achieving the same resolution. An example is a cross of spaced elements (Fig. 5a), a type of array introduced by Christiansen and Mathewson (1958) which uses the "cross" principle of Mills and Little (1953). In this case the signals received in the two arms are combined not by simple addition but rather by multiplication or *correlation*. Thus one obtains a power polar diagram which is the product of the voltage polar diagrams of the two separate arms: one arm gives a set of parallel, spaced, narrow ridges in the x direction, the other a similar set in the y direction, and so the product is a pattern which is nonzero only in the regions where the two sets of beams overlap, i.e., a matrix of pencil beams rather like that given by the matrix array. In the present case the cross sections of the voltage polar diagrams of the orthogonal ridges are $(\sin \pi a x)/\pi a x$ and $(\sin \pi a y)/\pi a y$, so that the shape of the power polar diagram of each pencil beam is $[(\sin \pi a x)/\pi a x] \ [(\sin \pi a y)/\pi a y]$ rather than the square of this quantity as obtained with the matrix array. The angular resolution is thus less than that of a matrix of the same overall size, and in addition the sidelobe level becomes undesirably high. However, sidelobes can be reduced to an acceptable level by tapering the current amplitude distribution along the array; when this has been done the resolution is half that of the matrix—in other words, each arm of the cross needs to be twice as long as the side of the matrix array for the same resolution.

It often helps to consider the performance of an array in terms of the Fourier transform $\bar{p}$ of the power polar diagram rather than the power polar diagram p itself. Previously we derived p as the Fourier transform of the aperture distribution u by Eq. (3). With the aid of the convolution theorem Eq. (3) may be written in the form

$$p = \overline{(u * u)}$$

or

$$\bar{p} = u * u, \tag{4}$$

where $*$ denotes the process of two-dimensional convolution. Given any array of spaced, small elements the $\bar{p}$ pattern can be simply drawn by plotting a point for each pair of elements at a distance from the origin corresponding to the spacing of that pair and in the direction of their

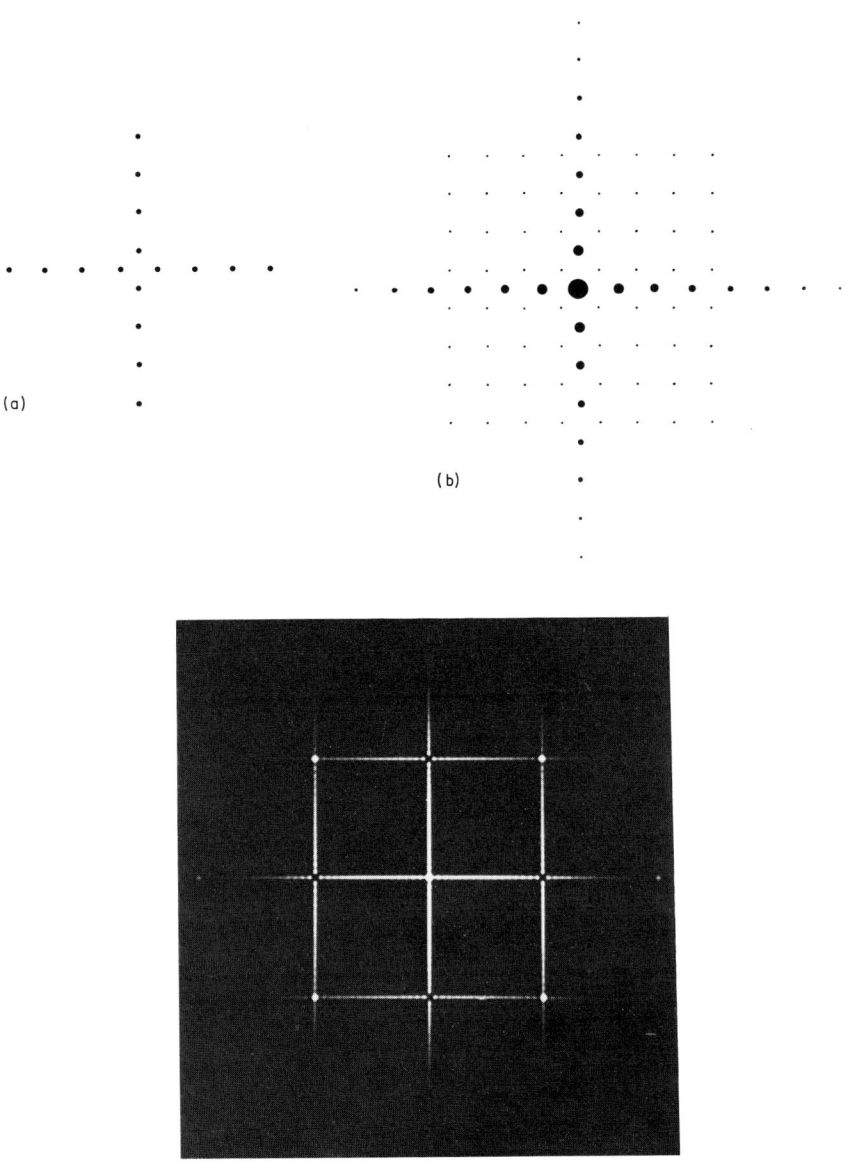

Fig. 5. (a) An 8×8 element crossed-grating array. (b) The autoconvolution function $u * u$ of this aperture distribution (see text). (c) The central part of the polar diagram of crossed-grating array, corresponding to the Fourier tranform of (b). As discussed in the text, the "fan beam" responses can readily be eliminated, leaving the pattern of pencil beams. With the array geometry shown in (a), alternate pencil-beam responses are of opposite signs.

separation. For example, the $\bar{p}$ pattern for a Christiansen spaced-element cross, regarded as a total power instrument, is as shown in Fig. 5b and the corresponding polar diagram in Fig. 5c. However, when the cross is operated as intended the signals in one arm A are correlated only with those in the other arm B; thus we calculate $\bar{p}$ (where p now means the product of the two voltage polar diagrams) using Eq. (4) in the modified form

$$\bar{p} = u_A * u_B$$

and so obtain a $\bar{p}$ pattern which is simply a square matrix of dots of uniform weight. Since $\bar{p}$ is the Fourier transform of the power polar diagram, the larger the area covered by this pattern the smaller the angular area of the beam.

The same pattern is obtained if half of one of the two arms of a cross is removed, resulting in a "T," the extra half arm of the cross being redundant.

This discussion of the cross array illustrates the geometrical requirement of a low-redundancy array; the configuration should be such that its autocorrelation pattern covers an area as extensive as possible in both dimensions with sufficiently frequent sampling throughout the area to ensure an adequate field of view to accommodate the solar image. There are doubtless many geometries which satisfy this requirement with varying shades of efficiency. However, apart from the cross or T, only one other geometry has been found in which the unwanted (or excessively weighted) Fourier components can be dealt with by direct analog means. This is the circular array (Fig. 8). In this case, a number of methods have been proposed for reweighting the Fourier components, the most satisfactory one being the process of "J^2-synthesis," which will be discussed in the section describing the Culgoora radioheliograph. The circular geometry is more compact than the cross by a factor of two, and its circular symmetry has certain advantages particularly for solar studies. On the other hand, the cross is more flexible for general radio astronomy, since it can also be employed in a fan-beam mode which has proved useful, for example, in the search for pulsars.

C. Processing the Received Signals

Having established a suitable geometry for the array aperture we now face the problem of how best to process the received signals in order to form an image in as short a time as possible. In the most general way one may visualize image production by supposing the interference pattern of each pair of antennas in turn to be displayed as a plane wave on an image tube and then adding all plane-wave components together photographically. If there are n separate antenna elements in our array then we may record n separate signals and, what is more important, $\frac{1}{2}n(n-1)$ cross-correlations (interference patterns). Some of these cross-correlations will be redundant (corresponding to the overweighted or unwanted Fourier components in Fig. 5b) and the number of "useful" independent cross-correlations will be $N = \frac{1}{2}\alpha n(n-1)$, where the redundancy factor α is typically $<\frac{1}{2}$. The maximum number of independent picture points in the image is then $2N$ (the factor 2 is included because each correlation is complex and so specified by two numbers). It is important to realize that the n signals contain information over the whole field of view determined by the spacing of the antenna elements. The apparent paradox (Gabor, 1969) that $\sim n^2$ picture points seem to be specified by only n signals is at once removed when it is realized that we are dealing with incoherent noise signals received in a finite frequency bandwidth Δf. At any one time our instantaneous "image" contains only n pieces of information, but during a finite time Δt we are provided with $\Delta f \Delta t$ such independent images, which in due course combine to provide the complete information. The final picture is thus the aggregate of a large number of rapidly changing sub-images.

One method of image-forming, which at first sight might appeal to the modern engineer, is to take all n signals, feed them into a computer which would rapidly execute the cross-correlations, perform the Fourier transforms, and so display the image with perfect efficiency. Unfortunately, however, logistics make this method impracticable, at least at the present time. The point is that since all n signals must be sampled Δf times per second the computer rates become excessive, e.g., with $n = 100$, $\Delta f = 1$ MHz the computer would receive 10^8 input numbers per second on which computations would have to be made. Nevertheless this "ultimate" technique may already be relevant to very-low-frequency studies where very narrow bandwidths are permissible. In the latter case one also has the option of recording on video tape for the subsequent processing, at leisure, of selected phenomena.

Since the prospects of completely digital processing are limited, one is driven to explore analog or hybrid methods. Such methods can be divided into two classes. In one method that has been proposed the radio signals are converted into some other kind of wave motion such as light or ultrasonic waves. An example is shown in Fig. 6, which is a suggested optical processor for a circular array. A collimated light beam is incident upon an aperture which transmits light through pupils that represent a scale model of the array. When this model is placed in the aperture of

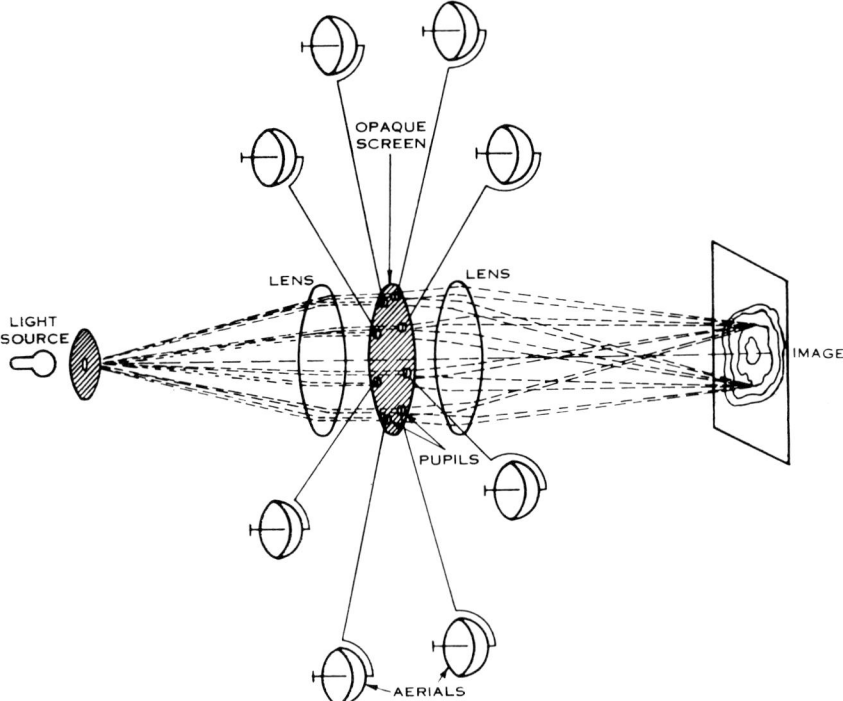

Fig. 6. A proposed optical processor for forming images from the signals received by a circular array of antennas (from McLean and Wild, 1961).

a lens the image is the Fraunhofer diffraction pattern of the aperture distribution. If the pupils were simple holes this pattern would correspond to the polar diagram of the array (basically a bright spot). However, the pupils are in fact designed to impose upon the transmitted light the phase and amplitude of the radio signals applied to them. A little consideration shows that the image on the screen then corresponds to the wanted radio image (convolved with the polar diagram of the array—the effects of wrongly weighted Fourier components must be removed by a further process). No radioheliograph has yet been made employing such concepts, though a detailed proposal has been made (McLean et al., 1967) and similar methods have been applied in other fields (Lambert et al., 1965; Briggs and Holmes, 1973).

In a second class of signal-processing method the required Fourier transform is performed by a branching network of cables. In principle the n signals from the aerials are converted to the $\sim n^2$ picture-point responses by interconnection through $\sim n^3$ carefully phased lengths of cable. This technique, introduced on a limited scale by Blum (1961), has not yet been fully exploited to the extent of providing simultaneous storage over a complete picture. The most extensive branching network so far used in a radio telescope, providing simultaneous storage over one complete picture *line*, is incorporated in the Culgoora radioheliograph (see Section V).

IV. The Evolution of the Radioheliograph

We have defined the radioheliograph as an instrument for rapidly producing radio images of the Sun. Aperture synthesis methods which require observations over extended periods are excluded by this definition; we should remark, however, that the first high-resolution quiet-Sun maps were obtained by aperture synthesis (O'Brien, 1953; Christiansen and Warburton, 1955).

The earliest radioheliographs employed fan-beam aerial systems, giving useful angular resolution in only one direction, with only moderately fast image formation. In this class were the first high-resolution grating arrays, which, as already mentioned, were built by Christiansen (1953), with $4'$ arc beamwidth at $\lambda = 25$ cm, and by Tanaka and Kakinuma (1954) with $4'.5$ beamwidth at $\lambda = 7.5$ cm. Both these instruments made use of scanning by Earth rotation to produce a one-dimensional image every 4 or 5 min.

A number of more sophisticated grating-array systems have since been built to provide the more rapid image formation needed for the recording of solar bursts. In several of these instruments the grating array is combined with one or more additional elements along the extension of the array baseline, to form a "compound interferometer" (Covington and Broten, 1957). This configuration gives a considerable improvement in angular resolution at small cost. For example, in the arrangement shown in Fig. 7 the Fourier components corresponding to all multiples of the array spacing b up to the length $AC_2 = 2nb$ can be obtained by recording correlations between the signals at each of the aerials C_1 and C_2 and the signals at each of the array elements. Thus the effective aperture is increased from $(n-1)b$ to $2nb$, and the angular resolution is approxi-

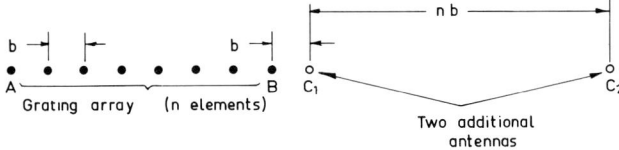

Fig. 7. A compound interferometer (see text).

mately doubled (for a chosen weighting of the Fourier components) by the addition of only two extra antennas.

High time resolution with one-dimensional angular resolution has been achieved by various methods. Tanaka et al. (1969) introduced rapidly varying phase shifts between the signals from the aerial elements to scan a fan beam across the Sun, so obtaining one image in 10 sec at $\lambda = 8$ cm with very high angular resolution (23″ arc beamwidth). The 169 MHz east–west array at Nançay, France (Vinokur, 1968) uses a branching network to produce multiple fan-beam responses; the signals are recorded photographically and digitally, giving images with 3′.8 beamwidths up to the high rate of 30 per second. In the case of the 160 MHz fan-beam radioheliograph at Nobeyama, Japan (Takakura et al., 1967), two separate compound interferometers give beamwidths 1′.8 east–west and 3′.4 north–south, respectively. Electronic phase sweeping is used to scan the Sun 10 times per second. A swept-frequency grating array at the Clark Lake Observatory, California, is used to observe both the spectrum and the east–west brightness distribution of the meter-wave solar radiation (Kundu et al., 1970). The receiving frequency is scanned from 20 to 60 MHz four times per second; during the scan the accompanying phase changes sweep successive lobes of the fan-beam response pattern across the Sun. In this way the instantaneous position of each emitting region is recorded during each scan for a series of points in the frequency range.

Digital techniques now offer an attractive alternative to these analog methods of rapid imaging. All the necessary cross-correlations between the signals from elements of the array are measured; a digital computer is then used to derive the brightness distribution by a real-time Fourier transform operation. This method is in use at Fleurs, Australia, for processing the signals from a 34-element compound interferometer. In this case, the east–west brightness distribution is recorded with 50″ angular resolution and with time resolution of a few milliseconds.

The two earliest pencil-beam radioheliographs were both crossed-grating arrays: at Fleurs, Australia (Christiansen and Mathewson, 1958) for $\lambda = 21$ cm, and at Stanford, California (Bracewell and Swarup, 1961) for $\lambda = 9.1$ cm. In both cases a combination of Earth rotation and phase steering was used to produce two-dimensional solar maps with about 3′ arc resolution at a slow rate (tens of minutes). The Fleurs instrument is at present being adapted for rapid solar imaging (W. N. Christiansen, private communication); in the new system, the necessary correlations will be measured and then combined with the aid of an on-line computer so as to form a two-dimensional pencil beam image in less than 1 sec.

A broadband "T" array now nearing completion at the Clark Lake

Observatory (Erickson, 1973) will be an extremely effective radioheliograph. This instrument, designed for use in both solar and galactic radio astronomy, will have pencil-beam resolution of a few minutes of arc at any frequency from 20 to 120 MHz. It will be capable of rapid image formation and multi-frequency operation; both the aerial beam pointing and the operating frequency will be controlled via a computer, and will be adjustable in less than 1 msec.

V. The Culgoora Radioheliograph

A. GENERAL DESCRIPTION

.This instrument uses circular arrays of antennas (Fig. 8) and produces high-resolution radio images of the Sun at the rate of one per second (higher picture rates—two or four per second—are available with reduced field of view). Two images, in right- and left-handed circular polarization, respectively, are built up during each picture period, rows of picture points being recorded for each image in turn. Three observing frequencies are used: 160, 80, and 43.25 MHz. The emission at these fre-

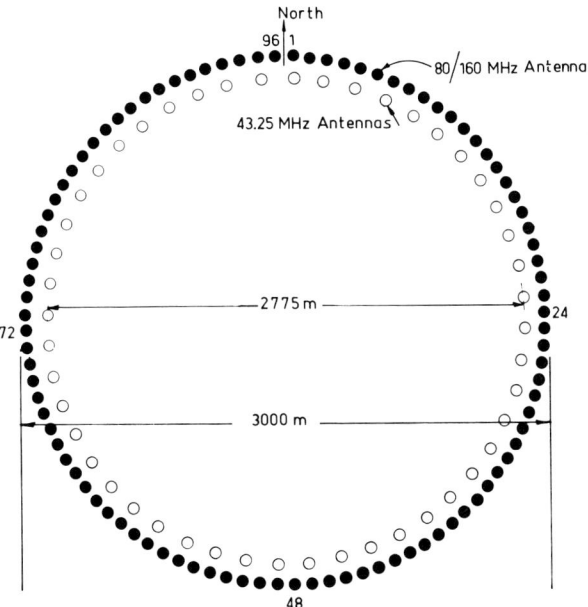

FIG. 8. The Culgoora radioheliograph arrays. The system of antenna numbering used in Section V is indicated in this diagram.

quencies is mainly generated at heights in the corona ranging from about 0.3 $R\odot$ above the photosphere at 160 MHz, where $R\odot$ is the photospheric radius ($R\odot \simeq 7 \times 10^5$ km), to 1 $R\odot$ at 43.25 MHz, so that the radioheliograph is well suited for radio studies of the middle layers of the corona. The frequency in use is automatically changed between 1 sec picture periods in a sequence selected by the operator.

An array of steerable parabolic reflector aerials with log-periodic dipole feeds is used at both 80 and 160 MHz. The array for 43.25 MHz reception is made up of 48 aerials of a much simpler corner-reflector design. The diameters of the arrays in terms of wavelength are in an exact 4:2:1 ratio, an arrangement which, as we shall see, considerably simplifies the phase-control system used for forming and pointing the aerial beams at all three frequencies. The radioheliograph commenced operation in September 1967, at first with only a single receiving frequency, 80 MHz (Wild, 1967).

B. Array Beam Formation

If the 96 elements of the 80 MHz array are connected together in phase the resulting power polar diagram is of the form shown in Fig. 9b. This picture was obtained by photographing the far-field diffraction pattern of a circle of 96 small holes in an opaque screen; as we saw when discussing electrooptical processors in Section III, C, this sytem is a precise optical analog of the antenna array. Figure 9a is a radial cross-section of the pattern.

The inner part of the polar diagram consists of a central maximum surrounded by decreasing concentric rings. The power response $F(r)$ at an angular distance r from the center of the pattern is given for array radius a and wavelength λ (Wild, 1961) by

$$F(r) = J_0^2(2\pi a r/\lambda), \qquad (5)$$

where $J_n(x)$ is the Bessel function of the first kind of order n. This is identical with the response of a continuous annular aperture of the same radius. However, the use of discrete elements to simulate an annulus leads to further responses at angular displacements from the central lobe which are integer multiples of r_1,

$$r_1 = n\lambda/2\pi a = \lambda/D, \qquad (6)$$

where n is the number of elements in the array, and D the distance between adjacent elements. These unwanted "grating" responses (the outer ring in Fig. 9b) are analogous to the first- and higher order responses of a diffraction grating.

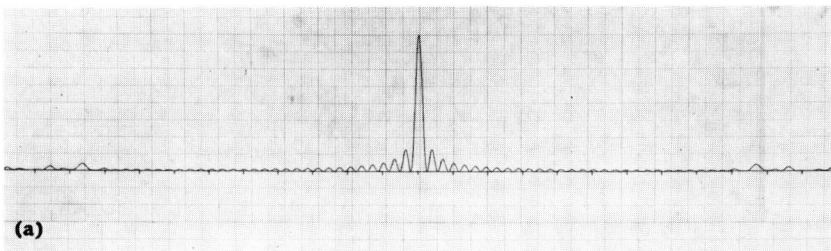

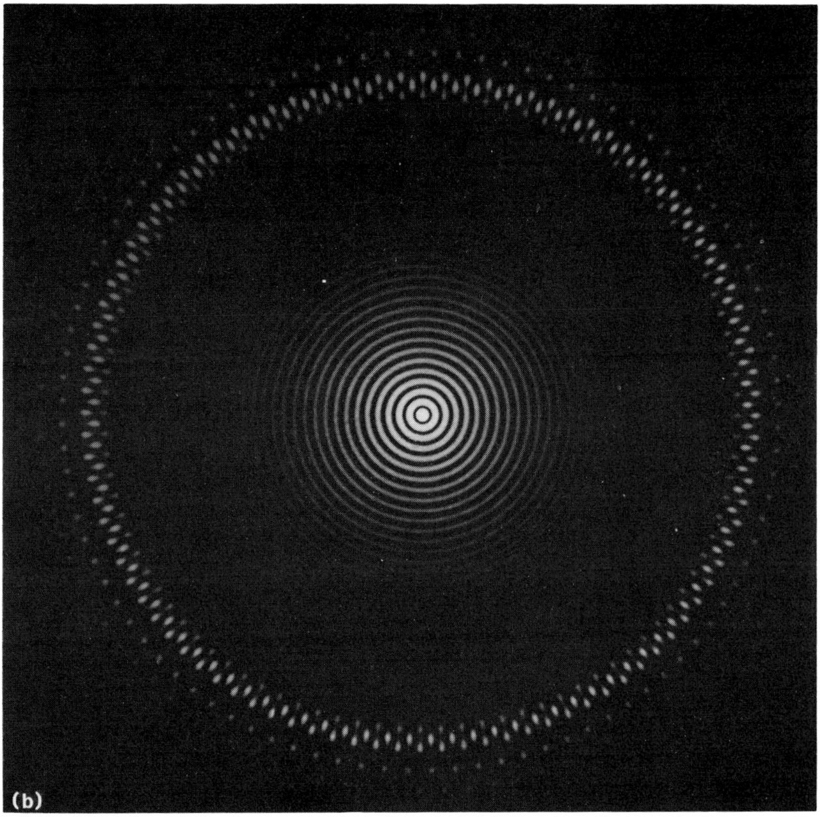

FIG. 9. The response of a 96-element circular array, with co-phased elements (after Wild, 1967, p. 283). (a) The computed radial cross section of the array pattern, showing the central field and the first-order grating response. (b) The optical diffraction pattern of an aperture consisting of a circle of 96 small holes. This pattern is an exact representation of the power polar diagram of a 96-element array.

Thus the polar diagram of the circular array has unwanted sidelobes of two kinds—the central rings and the grating responses. With discrete elements the latter cannot be eliminated; however, as discussed in Section III, their effects can be avoided by choosing the spacing D so that r_1 is large enough to ensure that grating sidelobes of the solar image always fall outside the required field of view.

The central ring pattern must be converted into a pencil beam with low sidelobes. Several methods of making the necessary changes in the weighting of the Fourier components of the response have been considered (Wild, 1961, 1965; Carter and Wild, 1964), and we shall now outline the theory of the "J^2-synthesis" method (Wild, 1965), which has been adopted for real-time polar diagram correction in the radioheliograph.

In addition to the $J_0^2[2\pi ar/\lambda]$ pattern which is produced by a co-phased array, other circularly symmetrical patterns can be generated by introducing progressive phase shifts from one antenna to the next round the circle. With uniform phase steps of $2\pi k/n$ between successive elements, i.e., $2\pi k$ round the circle, where k is an integer, the polar diagram is found to be

$$F(r) = J_k^2(2\pi ar/\lambda). \qquad (7)$$

Some of these patterns are plotted in Fig. 10a, which at once suggests the possibility of removing unwanted sidelobes by subtracting from the J_0^2 response some suitable combination of J_k^2 patterns (for example, $J_0^2 - J_2^2$ is a marked improvement on J_0^2 alone).

A complete analysis (Wild, 1965) confirms that this is indeed the case and that any polar diagram $F(r)$ within the resolution limit of the aperture can be synthesized by a suitable summation of J_k^2 patterns:

$$F(r) = \sum_{k=0}^{\infty} t_k J_k^2\left(\frac{2\pi ar}{\lambda}\right). \qquad (8)$$

With an n-element array, the meaningful range of k is $0 \leq k \leq n/2$, so that this infinite series can be replaced with a finite one—with 49 terms when $n = 96$. The number of terms can in this case be reduced to between 10 and 20 without appreciably affecting the result, and a series of 16 terms is used to generate the radioheliograph polar diagram (Fig. 10e).

In the radioheliograph the J^2-synthesis correction process is implemented by introducing phase shifts to produce each in turn of the response patterns $J_k^2(2\pi ar)$ for a time proportional to $|t_k|$. The resulting signals are integrated, giving the effect of the summation in Eq. (8). To take account of the fact that some of the coefficients are positive and others negative the polarity of the integrating process is reversed at ap-

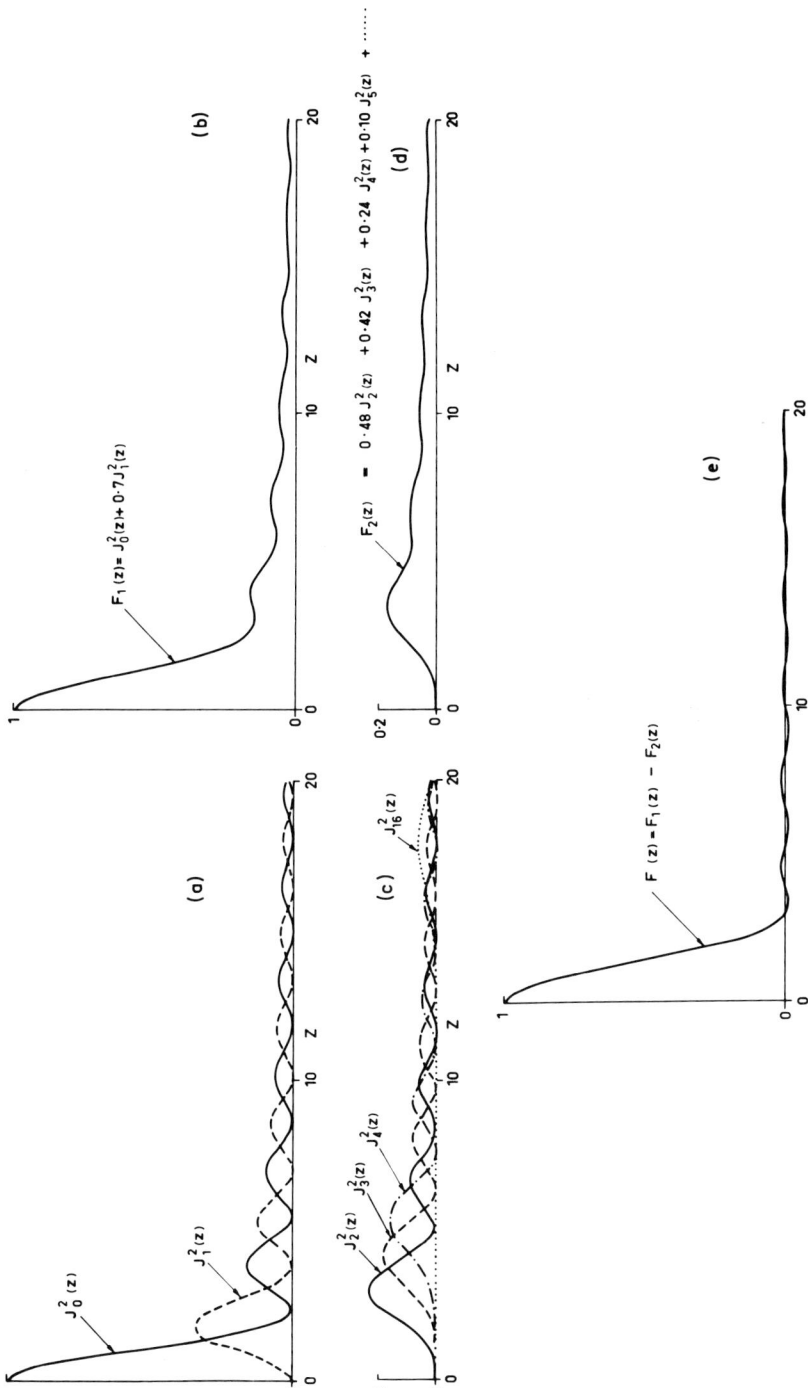

Fig. 10. An illustration of successive steps in the "J^2-synthesis" method of apodization for the polar diagram of the circular array (from Wild, 1967, p. 284).

propriate instants. This complete sequence of operations must, of course, be executed during the 8 msec period allowed for building up each picture point.

C. Picture Formation

With the array dimensions shown in Fig. 8 the field of view that is free from grating responses is a circle, of diameter 1° at 160 MHz and 2° at both 80 and 43.25 MHz. In practice it is much simpler to scan a rectangular rather than a circular field of view, and the picture formats used for the three frequencies are shown in Fig. 11. The picture-point spacing is in each case set as large as possible without losing information, as determined by the sampling theorem. The diagrams in Fig. 11 apply

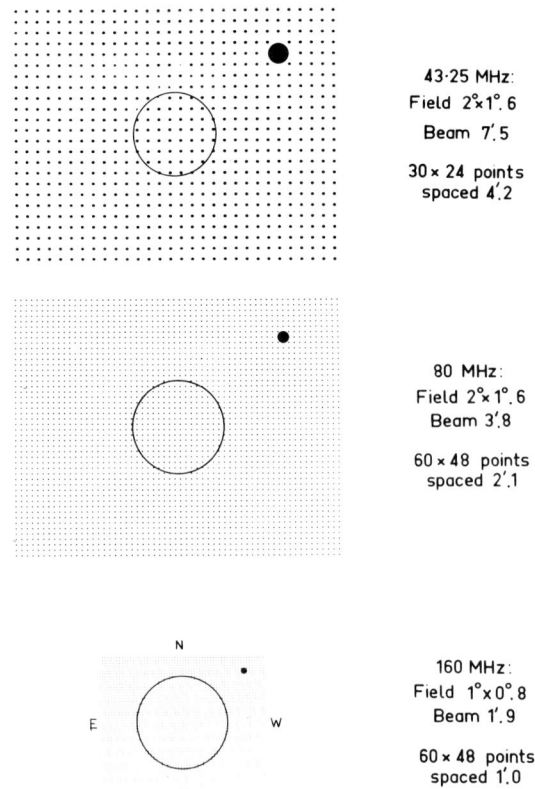

43·25 MHz:
Field 2°x1°.6
Beam 7'.5

30 x 24 points
spaced 4'.2

80 MHz:
Field 2°x 1°.6
Beam 3'.8

60 x 48 points
spaced 2'.1

160 MHz:
Field 1°x 0°.8
Beam 1'.9

60 x 48 points
spaced 1'.0

Fig. 11. The radioheliograph picture formats for the three operating frequencies. The large black dot shows the half-power contour of the power response in each case; the superimposed circles represent the position of the photospheric disk (from Sheridan et al., 1973).

to fields centered on the zenith; in other directions foreshortening effects broaden both the aerial beam and the field of view.

Because of the requirements for good sensitivity and rapid picture formation (see Section III, C), it is not practicable to build the picture by scanning a single beam to each of the $60 \times 48 = 2880$ picture points in turn. Instead, a branching network is used to form simultaneously a north–south line of pencil beams which are then scanned from east to west in steps equal to the required picture-point spacing. Midway through each step of the scan the mode of reception is changed from right- to left-handed circular polarization; thus the information for each point in either of the two pictures is received for about 8 msec.

D. Array Beam Steering

To direct the beam of the radioheliograph toward a point $(\Omega t, \delta)$ on the celestial sphere (Ωt and δ are the equatorial coordinates, hour angle and declination, respectively, of the selected point, t being the apparent solar time) we must add or subtract a path length in each aerial channel. The excess path required in the jth channel is given by

$$P_j = a\{\sin \alpha_j \cos \delta \cos \Omega t - \cos \alpha_j[(1 - \cos \Omega t) \cos \delta \sin L - \sin (L - \delta)]\}, \quad (9)$$

where L is the latitude and α_j is the azimuth (measured eastward from north) of the jth aerial with reference to the center of the array. In practice, the path differences P_j are simulated by introducing into each channel a phase shift

$$\phi_j = 2\pi[(P_j/\lambda) - m] \quad (10)$$

where m is an integer such that $0 \leq \phi_j < 2\pi$.

In Eq. (9), the sin α_j and cos α_j terms correspond respectively to the eastward and northward components of the separation of the jth aerial from the center of the array. For the real-time phase-shift computations (Section V, F) it is convenient to separate these "east" and "north" components. The computer logic is further simplified by deriving separately the slowly varying "tracking" phases T_j, which keep the field of view centered on the Sun, and the rapidly changing "scanning" phases S_j, which move the pencil beams east-to-west across the field in each picture period:

$$\phi_j = T_j + S_j. \quad (11)$$

Here, T_j is the value of ϕ, from Eqs. (9) and (10), which at the start of the picture period aligns the row of pencil beams with the eastern edge of the required field. The addition of the S_j phase components moves the beams step-by-step westward across the field of view to build up the picture:

$$S_j = \left(\frac{2a \sin \theta}{\lambda}\right) N(t) \sin \alpha_j,\tag{12}$$

where $N(t)$ is a "staircase" function with unit steps and θ is the angular picture-point spacing at zenith.

E. The Receiving System

This is shown in simplified block-diagram form in Fig. 12. The 80 and 160 MHz signals at each aerial pass first through polarization switches, which select right- and left-handed circular polarization during alternate picture-point periods (polarization is not recorded at 43.25 MHz). The signals are then transmitted on open-wire lines to the observatory building at the center of the arrays. Here the 160 and 43.25 MHz signals are first converted to 80 MHz by heterodyne mixing. The signals for the frequency in use are then selected by electronic switches and converted to the 7 MHz intermediate frequency at a second mixer. The computer-controlled phase shifts required for the formation and steering of the aerial beams are applied at this point by phase switches in the local oscillator feed lines. Each of these phase switches uses a binary arrangement of seven sections giving phase steps of $2\pi/2, 2\pi/4, \ldots, 2\pi/128$,

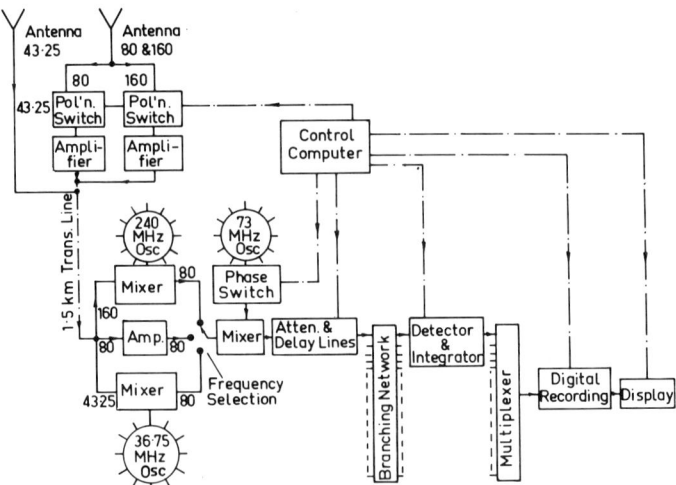

Fig. 12. Block diagram of the radioheliograph receiving system. Linkages to the control computer are indicated by dash-dot lines. Details for one channel are shown with the other channels indicated by multiple connections to oscillators, branches, network, etc.

respectively, so that any phase shift from 0 to 2π can be produced, in steps of $2\pi/128$.

In the subsequent IF stages provision is made for the insertion (also controlled by the computer) of variable lengths of delay line in each of the 96 channels. The required delay follows the value of P_j [Eq. (9)] in coarse steps of one IF wavelength. In this way the signal paths from the source via each of the aerials to the point where the signals are combined are approximately equalized to preserve coherence over the 1 MHz bandwidth. The IF also incorporates an automatic gain control, to ensure that the output signal remains within the dynamic range of the receiving and recording system, even during intense and rapidly varying bursts.

The 96 incoming signals are combined in the branching network to give 48 outputs which correspond at each instant to one column of picture points. These output signals are rectified, and integrated over each 8 msec picture-point period.

The integrated dc output voltages are transferred to a buffer, and during the next 8 msec period are read serially by a multiplex sampling unit (at the same time the integrators are building up output voltages for the next set of picture points). The sampled values are converted to digital form and recorded on magnetic tape (see Section VI).

Immediately after the record has been written on the tape, the latter passes over a "read" head, the signals from which are decoded and fed to a video display system. In this way a pair of real-time cathode ray tube (CRT) pictures of the radio Sun (for the two senses of circular polarization) is obtained. The deflection voltages for the cathode ray tubes are automatically set to preserve the scale of the picture (see Fig. 11) as the operating frequency is changed second by second. In addition, since both the size and shape of the area of sky included in the field of view change with the position of the Sun, a further set of corrections is superimposed on the deflection voltages so that an undistorted picture with uniform scale is presented at all times.

F. THE PHASE-CONTROL SYSTEM

1. The Control Computer

An important aspect of the radioheliograph design is the method adopted for real-time control of the complicated signal-processing sequence. The instrument was designed at a time when digital techniques were only just coming into widespread use in radio astronomy and analog methods were still the more familiar approach. For instance, it is interesting to recall that one of the early proposals for phase control involved

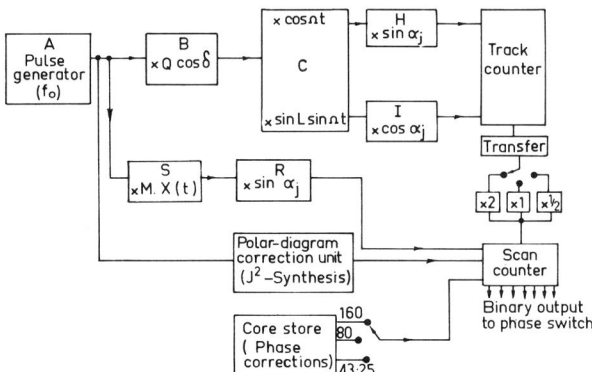

FIG. 13. Block diagram of the phase-control section of the computer control.

the use of mechanical phase shifters, operated by a formidable gear train with 96 outputs! The system finally adopted, however, uses digital techniques for both control and data handling; indeed, the required accuracy and high operating speed could not have been attained by analog methods.

The nerve center of the system is a special-purpose digital computer, which generates all the controlling waveforms for the receiving, recording, and picture-making circuits of the radioheliograph, as indicated diagrammatically in Fig. 12. Its most demanding function is the alignment in phase of the signals from the 96 aerials before these signals are combined in the branching network.

Figure 13 shows the organization of the phase-control section of the computer. The binary phase switches (see Fig. 12) are driven directly by the corresponding stages of 7-bit "scan counters," which form the output of the computing system. The unit of phase change in these switches is $2\pi/128$, or about $2°.8$, corresponding to a path change of about 3 cm at 80 MHz frequency. Equation (9) shows that P_j can be of the order of 1 km, so that an overall system accuracy of about one part in 5×10^4 is required. The contents of each scan counter (interpreted as a binary number) must at any time be equal to the sum of T_j and S_j [Eqs. (9)–(12)] plus the J^2-synthesis component. Accumulated counts in excess of 128 units ($\equiv 2\pi$ rad) overflow and are lost, thus automatically accomplishing the subtraction of the $2\pi m$ term in Eq. (10).

2. Computation of Tracking Phase

The slowly varying counts for T_j are stored in a set of "tracking counters," which are fed with pulses at a rate dT_j/dt, derived from Eqs.

(9)–(11). This is achieved by appropriately modifying the flow of pulses from an oscillator A which runs at a constant frequency f_0 (Fig. 13). The remaining stages are "pulse-rate converters," which reduce the pulse rates by factors corresponding to the δ-, t-, and j-dependent terms in Eq. (9). At the output of unit B the rate is $f_0 Q \cos \delta$, where Q is given by

$$Q = \frac{2\pi a \Omega}{\lambda \phi_1 f_0} \qquad (13)$$

and $\phi_1 = 2\pi/128$ is the unit phase step in radians. The conversion rate $Q \cos \delta$ is set daily according to the solar declination, by means of manual switches.

The second converter C has two output networks to give conversion rates related to the east–west and north–south components of tracking phase. The conversion rates are $\cos \Omega t$ and $\sin L \sin \Omega t$, respectively, and are adjusted automatically at 5 min intervals to give the required variation of pulse rate with time.

The final stage of conversion is also carried out separately for the east–west and north–south components. Each of the converters H and I has 24 outputs, with conversion rates, respectively, $\sin \alpha_j$ and $\cos \alpha_j$ ($j = 1$ to 24).

The result of these operations is to produce pulse rates of $f_0 Q \cos \delta \cos \Omega t \sin \alpha_j$ from the H unit, and $f_0 Q \cos \delta \sin L \sin \Omega t \cos \alpha_j$ from the I unit. For the channels $j = 1$ to $j = 24$ the H and I outputs are added, so that after integration for a time t the contents of the jth track counter are

$$C_j = f_0 Q \{\cos \delta \sin \Omega t \sin \alpha_j + \cos \delta \sin L (1 - \cos \Omega t) \cos \alpha_j\}. \qquad (14)$$

Because of the quadrant symmetries of sine and cosine, C_j can be obtained for channels corresponding to antennas in the other three quadrants of the circle by appropriate changes in the signs of the H and I outputs before addition and integration. The tracking counters are up–down counters to cope with both positive and negative inputs.

The initial conditions are fixed by clearing the counters and setting the computer to zero hour angle ($t = 0$) so that $C = 0$. For any hour angle the count C_j corresponds to the tracking phase shift T_j, except for a time-independent term $f_0 Q \cos \alpha_j \sin (L - \delta)$. This constant for the day is inserted by feeding in at the input of unit I a number of pulses equal to $f_0 Q \sin (L - \delta)$, a value which is set by manual switches to correspond to the Sun's declination. After this simple setting-up procedure, the computer is run to the hour angle corresponding to the time at which observations are to begin (in order to run backward in hour angle the signs of the inputs to the tracking counters are reversed).

3. Scanning

At the beginning of each 1 sec picture period the scanning counters are cleared. Each is then loaded with the contents of the corresponding tracking counter, to which is added a "calibration" count from a magnetic core store. The latter compensates for unavoidable phase-shift variations among the 96 signal channels; the stored numbers are determined from phase measurements on the system made every few weeks at each of the three frequencies. At this stage the 48 pencil beams are aligned with the eastern edge of the required field of view.

Two pulse-rate converters (units R and S, Fig. 13) feed pulses to each channel at a rate corresponding to the time derivative of the expression for S_j in Eq. (12); M in Fig. 13 is the constant factor $2a \sin \theta/\lambda$. These pulses are fed into the scanning counters at 16 msec intervals to move the pencil beams in steps westward across the field so that each column of picture points is recorded in turn. During each picture-point period a further more rapid train of pulses from a "polar diagram control unit" is added into the scan counters to provide the sequence of phase progressions for the J^2-synthesis process.

4. Multifrequency Operation

Up to this point we have considered the sequence in which the digital phase control performs the tracking, scanning, and beam-forming operations for repeated images at a single operating frequency. When the frequency is changed, a change in all tracking phases is needed in order to keep the picture centered on the same point in the sky. On the other hand, it is convenient to use the same set of scanning phase shifts for all frequencies, for in this way the appropriate angular picture-point spacing is maintained. In addition, no alteration is required in the sequence of phase shifts for J^2-synthesis.

Since the aperture in wavelengths is chosen to be in an exact $1:2:4$ ratio at the three frequencies, it follows from Eq. (10) that $1:2:4$ changes are also needed in the numbers transferred from the tracking to the scanning counters. These changes can be made very simply by merely shifting the digits of the binary number in the tracking counter before the transfer takes place. The arrangement used is indicated in Fig. 13; each transfer is made via a set of gates controlled by the frequency-programming unit.

VI. Culgoora Data Processing

Figure 14 illustrates the way in which data progresses from the original observations of the Sun made at the observatory at Culgoora until it is presented in a format to suit the purposes of the radio astronomer in Sydney. In the top "box," representing processing done at Culgoora, two paths are shown. The older procedure, soon to be phased out, is shown at the right; it consists of recording on magnetic tape and later making an edited copy of selected sections of the record. The new procedure involves real-time processing by a small computer to select which data are to be kept. The advantages of this real-time processing and the methods employed are discussed in Section VI, B.

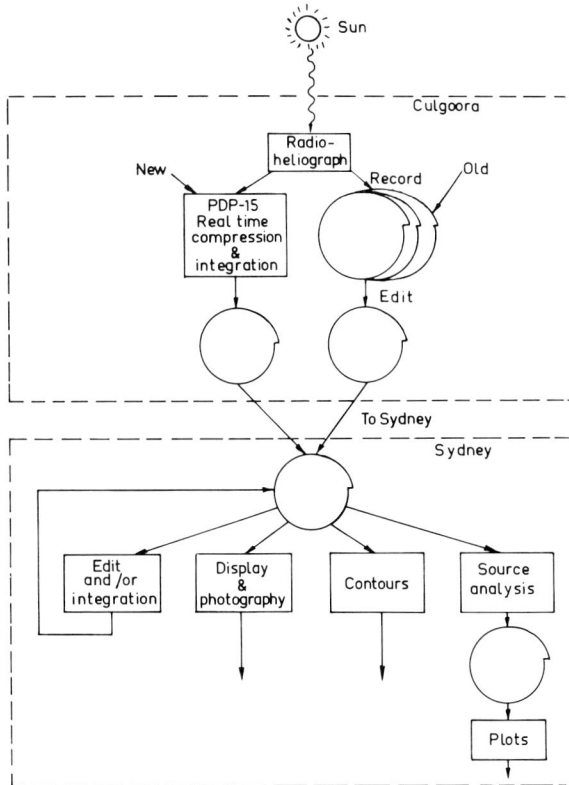

FIG. 14. Diagram illustrating the flow of radioheliograph data from the original observations to the presentation of the results in various forms suitable for analysis and publication. See text for details.

The lower box in Fig. 14 represents the rest of the processing of radioheliograph data, which is done in Sydney. This off-line processing is the subject of Section VI, A.

A. OFF-LINE PROCESSING OF CULGOORA DATA

1. The Data-Handling Problem

The data from the Culgoora radioheliograph are recorded digitally on "industry-compatible," seven-track magnetic tapes. At present the observer edits the data. On a day of average activity he discards some 95% of the day's record leaving about one 2400-ft spool of tape to be kept for subsequent analysis. (The new on-line data-handling system described in the next section will compress the whole day's record—some 24,000 pairs of images—on to about the same amount of tape.)

The fact that such a large quantity of digitally recorded data is accumulated for subsequent processing has naturally led to the extensive use of computers to manipulate the data. The manipulations are generally very simple; indeed they would be almost trivial but for the bulk of data involved. In the following sections it will be seen that the emphasis is on the presentation of data in more intelligible forms for study or publication, rather than on any mathematical transformation or attempted interpretation. If the reader feels that these manipulations are not very important he is invited to consider as he reads further how the radio astronomer, confronted by a great variety of solar phenomena, would be hampered if he were limited to one form of presentation (say, photographic film) which is not necessarily optimum for any of his purposes.

2. Editing for Filming

The edited tapes are played back on the CRT display, and each pair of pictures (together with a heading block showing date, time, observing frequency, and various receiver settings) is photographed on 35 mm film. This film record provides a very direct and convenient method of examining radioheliograph data. Figure 15 is part of a single frame showing a radioheliogram in one polarization.

When results of the highest quality are required, computer editing of the data is carried out before photography. This procedure is made necessary by the enormous variations both in intensity and time scale of the solar phenomena recorded. The brightness of solar burst sources varies from just high enough to be discernible above the quiet Sun level to about one million times this level. Some of these variations are very

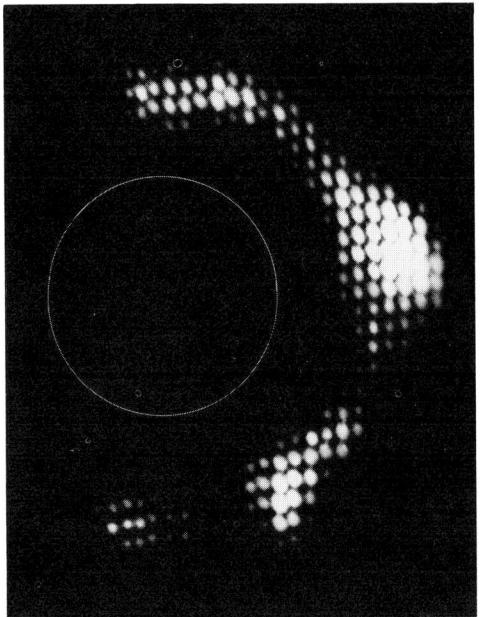

Fig. 15. Example of a frame of radioheliograph data. Each second two images of a field centered on the Sun are recorded in opposite senses of circular polarization. About half of one of these images is shown. The superimposed circle on this and other reproductions of radioheliograms indicates the solar photosphere.

rapid; a factor of 10 or 100 in a few seconds is not uncommon. The automatic gain control system uses attenuators which are switched into the IF channels in steps of 7 dB, but the residual variation is still greater than is acceptable for satisfactory filming. Furthermore, these residual variations are most misleading; each time the real brightness of a burst rises through a certain threshold its apparent brightness decreases suddenly owing to the insertion of an extra 7 dB attenuation. When the burst fades its apparent brightness leaps up as the attenuation is removed.

The obvious solution to this problem is to vary the video gain of the display from which the records are filmed to ensure correct exposure throughout. Since the changes are too sudden for manual control the only really satisfactory solution is to have the computer examine each frame and renormalize the data values or, with suitable hardware, set the video gain to accommodate the brightest feature. The faintest feature which can be recorded correctly is limited by the sidelobes of the brightest sources, and so this procedure does not cause any loss of information.

A second reason for using computer editing is to cope with the problem of variable time scale. Bursts occur with a great variety of time scales: some vary significantly in a second, some in a minute; others change appreciably only over a period of an hour or more. The heliograph recording rate of one image per second is designed to cope with all but the very fastest of these. However, a rapidly varying burst, recorded at 1 frame per second and played back at the normal projection rate of 24 frames per second will be over before the viewer realizes it has started. At the other end of the scale, one hour's records viewed at 24 frames per second takes $2\frac{1}{2}$ minutes—rather a long time to wait for something to happen. In the first case the solution is to slow the action by repeating each frame a number of times (24, if the original time scale is to be maintained). Conversely, slow changes become more obvious if the data are thinned. Both these manipulations are very simple to program in a computer.

An editing program has been developed which reads data from the original tape, multiplies all the data points in the image by a factor chosen to make the brightest point equal to some constant value, and writes the renormalized record on a new tape. This program also includes facilities for selecting which sequences of data are to be processed and for thinning or repeating frames to adjust the time scale.

3. *Contour Plotting*

Although the halftone display of data in the form shown in Fig. 15 has advantages, particularly for projection as a movie film, there are other cases where contours of brightness are more useful.

For example, a common problem is the comparison of the positions, shapes, and sizes of sources at different times, at different frequencies of observation, or in different polarizations. Using halftones this can be done very successfully by superposing two images, but only if one is prepared to go to the expense and trouble of using color to distinguish them. When this is not justified or not possible it is a relatively straightforward matter to superpose two or more contour plots, as for example in Fig. 16.

The data are sampled at the intersections of a raster which is skew, although uniform. The computer programming required for plotting contours from data in this form is quite simple, especially as in this case we can be satisfied with linear interpolation between points on the raster. The only sophistication incorporated in the program developed for this

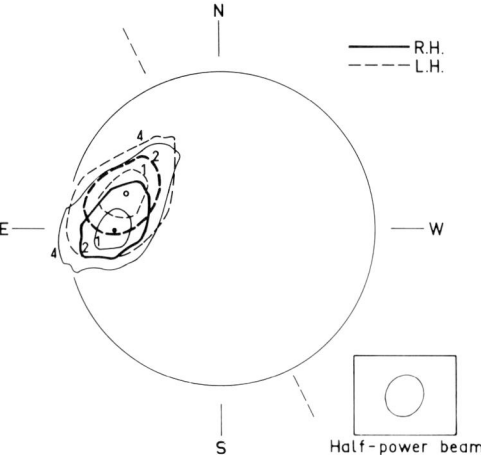

Fig. 16. Contours of a type I storm source observed in two senses of circular polarization, superimposed to show the relative displacement of the two source positions (from Kai, 1970).

purpose arose from the need to handle tapes efficiently, detect tape errors, determine the zero level, and provide general flexibility of control.

4. *Integration*

The radioheliograph was designed primarily for the observation of radio bursts, with adequate sensitivity for recording the emission from these sources at the specified picture rate. Its high resolution makes the radioheliograph eminently suitable also for mapping the quiet Sun; in this case, however, the source is much more extended than a burst source and so* a much longer integration time—about 100 sec—is required to give an image with adequate signal-to-noise ratio. This integration could be achieved by hardware if we had prior knowledge of when the bursts would occur and when the Sun would be quiet. Since this is impossible the solution adopted is to observe continuously at the fast rate necessary for bursts and later to sum selected groups of about 100 quiet 1 sec records in the computer to improve the signal-to-noise ratio. This procedure is a good example of the flexibility gained by the use of digital recording.

Examples are shown in Fig. 17 of quiet-Sun contour maps derived in this way at 80 and 160 MHz. The slight irregularities are produced

* With dilute high-resolution arrays, all parts of an extended source contribute to the noise, and the signal-to-noise ratio is consequently lower than it would be for a compact source of the same brightness (Wild, 1967, p. 289).

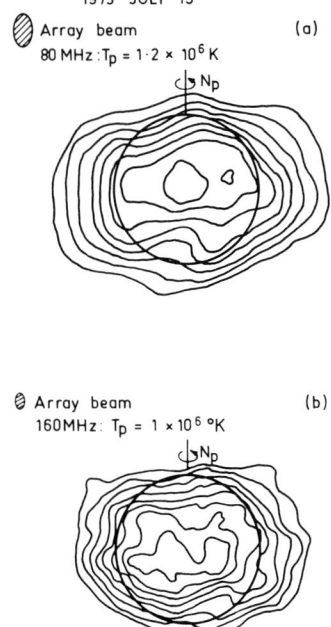

Fig. 17. Radioheliograph contour maps of the quiet Sun at (a) 80 MHz (b) 160 MHz. The contour levels are 0.9, 0.8, . . . , 0.1 times T_p, the peak beam brightness temperature, in each case. The half-power outline of the antenna beam is shown on each map; the north pole of the Sun is at the top of the diagrams.

by stable regions of slightly enhanced radio brightness which have been found to occur in association with active regions on the Sun.

5. Source Analysis

None of the processes discussed so far makes full use of the potential of digital recording to yield numerical data on such parameters as the degree of polarization, flux density, and size of the sources observed.

Since the vast majority of radio sources observed by the radioheliograph can be adequately described as bright blobs, most of the information in the records is retained if the observed distribution is represented by a set of elliptical Gaussian distributions, one for each resolved source on the original record. There are, of course, exceptions, such as the semicircular source illustrated in Fig. 15, but these are too rare to justify any attempt to allow for more complex source shapes in the program to be described.

The problem of analyzing a record into separate sources is of some

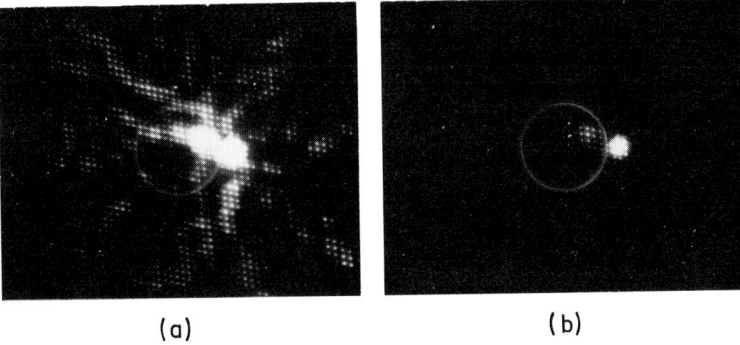

FIG. 18. (a) A reproduction of a heliograph record, grossly overexposed so as to emphasize the sidelobes. (b) The same frame, correctly exposed, shows only the two bright sources.

interest. First, it must be noted that associated with every source there are inevitably sidelobes. These are low-level brightenings of instrumental origin. Figure 18a shows a normal record heavily overexposed to accentuate the sidelobes, while in Fig. 18b the same record is shown correctly exposed. This example illustrates that the sidelobes are of relatively low level (normally the strongest sidelobes are about 5% of the main beam). Because of the sidelobes it is not possible to analyze sources which are weaker than a few percent of the brightest point on the Sun.

A second complication is the fact that solar sources are often double or multiple and it is desirable that the program should recognize their multiplicity. A third consideration is that the amount of data collected each day is large, and the funds for its analysis are finite, so that the amount of time spent analyzing each record cannot be too great.

In the method adopted the computer has been programmed to execute the following steps for each polarization separately.

(1) The brightest point of the image is found. The brightness of this point multiplied by the sidelobe ratio sets a threshold level. Points fainter than the threshold are not included in the analysis.

(2) The local maxima of the image are found and each maximum is tentatively assigned to a separate source.

(3) Points brighter than the threshold and adjacent to a point which has already been assigned to a source are also assigned to that source. Points are assigned in order of decreasing brightness. When all the sources in an image are distinct this process continues to completion without any complication.

(4) When two or more sources overlap there will be points brighter

than the threshold for which step (3) cannot be performed unambiguously because they are adjacent to points assigned to two different sources. When this situation arises the alternatives are (a) if the brightness of the point to be assigned indicates that the saddle point between the two sources is not very deep, then the two sources are merged and treated as one; (b) otherwise, somewhat arbitrarily, the ambiguous point is assigned to the same source as the brightest adjacent point.

(5) When all points above the theshold have been assigned, the following quantities are evaluated for each source:

$$F = \Sigma b_i$$
$$\bar{x} = \frac{1}{F}\Sigma b_i x_i \qquad \bar{y} = \frac{1}{F}\Sigma b_i y_i$$
$$\overline{x^2} = \frac{1}{F}\Sigma b_i(x_i - \bar{x})^2 \qquad \overline{xy} = \frac{1}{F}\Sigma b_i(x_i - \bar{x})(y_i - \bar{y}) \qquad (15)$$
$$\overline{y^2} = \frac{1}{F}\Sigma b_i(y_i - \bar{y})^2$$

Here b_i, x_i, and y_i are the brightness (beam flux density) and x and y coordinates of the ith point, and the summation is over all points assigned to a source. F is the source flux density and $\bar{x}$, $\bar{y}$ are the coordinates of the centroid of the source. An elliptical Gaussian brightness distribution is assumed for the source and the half-power major and minor diameters, D and d, and the orientation θ of the major axis are then determined from $\overline{x^2}$, $\overline{xy}$, and $\overline{y^2}$ by using the formulas:

$$\begin{aligned}\theta &= \tfrac{1}{2}\tan^{-1}\{2\overline{xy}/(\overline{x^2} - \overline{y^2})\} \\ D &= 2.36\{\overline{x^2}\cos^2\theta + 2\overline{xy}\cos\theta\sin\theta + \overline{y^2}\sin^2\theta\}^{1/2} \\ d &= 2.36\{\overline{x^2}\sin^2\theta - 2\overline{xy}\cos\theta\sin\theta + \overline{y^2}\cos^2\theta\}^{1/2}\end{aligned} \qquad (16)$$

The coordinates $x_{\max}$, $y_{\max}$, and brightness $b_{\max}$, of the brightness maximum are also recorded, and, by comparison with the image recorded in the other polarization, the degree of polarization $p_{\max}$ at the point of maximum brightness is determined and recorded. All these quantities must be transformed from the skew coordinate system in which the heliograph data is recorded to a rectangular system.

Finally then, for each source found by the computer and for each polarization the quantities computed are F, $\bar{x}$, $\bar{y}$, D, d, θ, $x_{\max}$, $y_{\max}$, $b_{\max}$, and $p_{\max}$.

The major fault in this scheme is that the same source seen in different positions relative to the raster will give slightly different results. In principle this could be overcome by an elaborate interpolation scheme, but the extra expense could scarcely be justified by the small gain in accuracy. An exception is in the estimated coordinates x_{max}, y_{max} of the brightness maximum. Obviously, without any interpolation at all these can only change in steps of the raster spacing; a very simple parabolic interpolation through the points bracketing the maximum, applied in each dimension separately, gives sufficiently improved accuracy.

Since all the above parameters are determined for each source present in each second of data, the output from this program still represents an overabundance of information. The results are recorded on a new magnetic tape ready for the second stage of the program, which is concerned with the presentation of the data in a variety of compact forms.

The flexibility of this second program is best illustrated by examples. Figure 19a shows a plot in which the brightness of some very rapid bursts has been plotted as a function of time. In this case an expanded time scale was essential for clear illustration of the rapid changes. By contrast Fig. 19b shows several parameters of a long-lived storm. Several hours of data are included here and the time scale has been compressed severely to emphasize the slow variations. The user has complete control over the parameters he wishes to see plotted, and the layout of the plot (scales, origins, range of values covered, etc.). The examples in Fig. 19 show how the program can be used in each case to bring out clearly the most significant features of the data.

Another form of presentation which is often useful is a map of the successive positions occupied by the various sources during a period of activity. Figure 20 is a map produced from the same data used for Fig. 19b. In this case the computer was asked to indicate the flux density from each source by the size of the blob drawn at each position, and successive positions were joined. In order to avoid cluttering this diagram the data were first thinned rather severely. Another type of a map presen-

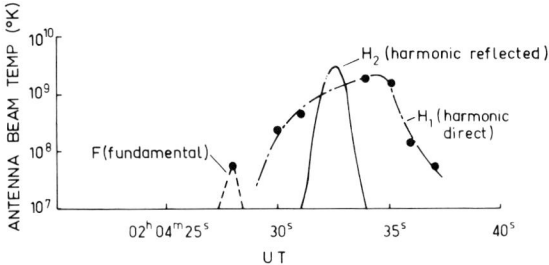

FIG. 19a. An expanded plot of antenna beam temperature of three sources observed during a short-lived type III burst (from Labrum, 1971).

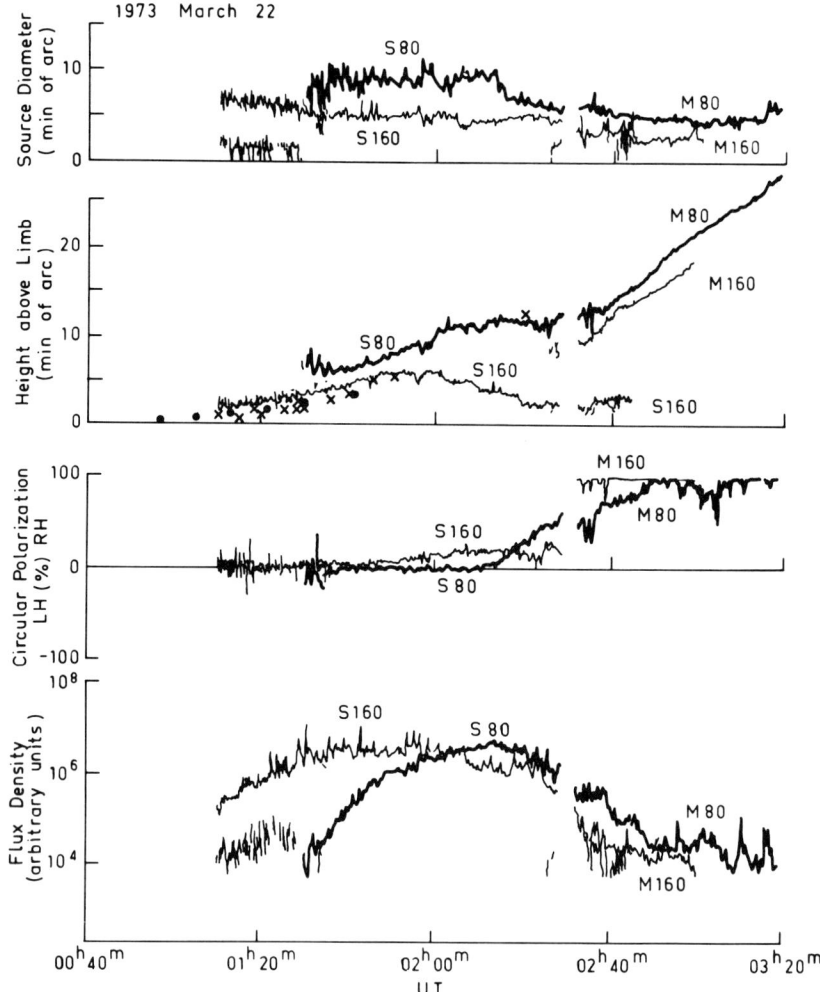

Fig. 19b. A plot on a compressed time scale of a number of significant properties of radio sources observed during a type I storm (labeled S) and moving type IV burst (labeled M). Figures on the curves denote the observing frequency (80 or 160 MHz) (from McLean, 1973).

tation is illustrated by Fig. 21. In this case the centroid of each source position during several hours of data has been indicated by a small mark, irrespective of its flux. Maps plotted by the computer for each polarization separately were combined photographically to form a composite in which the positions of sources seen in right-handed circular polarization

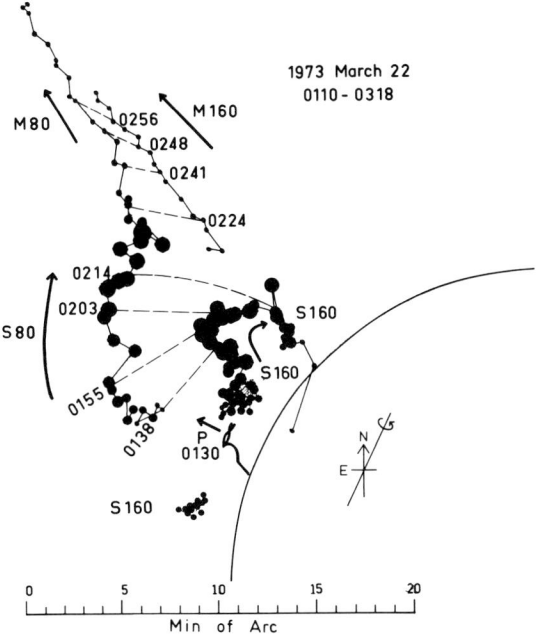

Fig. 20. A plot of the path followed by the moving source shown in Fig. 18b. Simultaneous observations at the two observing frequencies are joined by dashed lines; successive positions at one frequency are joined by full lines. The area of the dot marking each source position is proportional to the intensity of the source at that time and place (from McLean, 1973).

appear black and those seen in left-handed circular polarization appear white. The bipolar nature of these sources is thereby clearly indicated.

Another feature of this program which is sometimes useful is the inclusion of a method for correcting for ionospheric refraction. Figure 22a shows a plot of the variations of the apparent positions of two storm sources observed during winter at a time when ionospheric refraction was particularly bad. In this case the apparent movements are clearly ionospheric in origin, because the two sources, about 1 $R_\odot$ apart on the Sun, moved without any change of their relative disposition. To correct for this effect the position of one of the sources (A in the figure) was taken to be constant and the apparent shifts of source A were then subtracted from the positions of all the other sources. The resulting, corrected plot is shown in Fig. 22b, from which it is clear that the position of B was also stable and that the short-lived source C moved in a north-westerly direction on the solar disk.

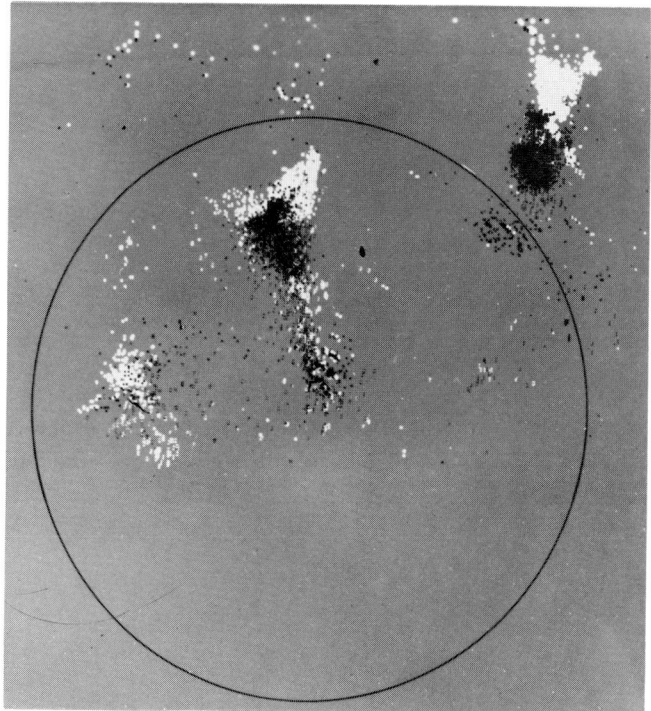

Fig. 21. Summary of position data taken from all available radioheliograms for November 19, 1970. The centroid position of each source for each second of data has been marked with a small dot—black for observations in right-handed circular polarization, white for left-handed (from McLean and Sheridan, 1972).

B. On-Line Data Handling

1. Data Compression and Integration

The recording of a day's observations with the Culgoora radioheliograph in the original format requires about 60,000 feet of magnetic tape. Even if the cost were not prohibitive the problems of storage of so much tape and of subsequent access would be.

At the time when the instrument was designed, experience with other less sensitive instruments suggested that it would be quite easy to cope with this problem by selecting the rare periods of activity and copying records to other tapes. However, the statistics of burst activity are quite different with the much more sensitive 96-aerial radioheliograph array, and it is true to say that periods of no activity are rare, except near sunspot minimum.

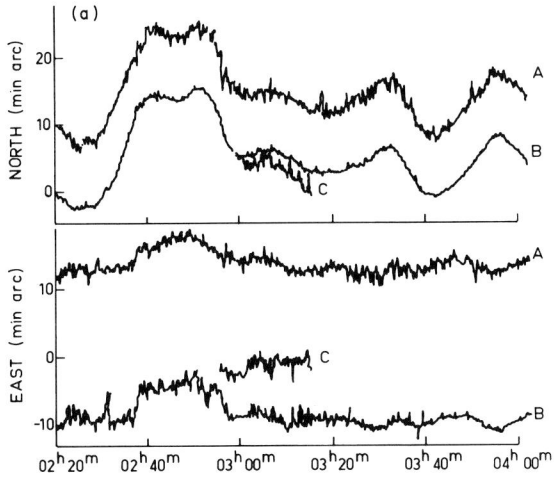
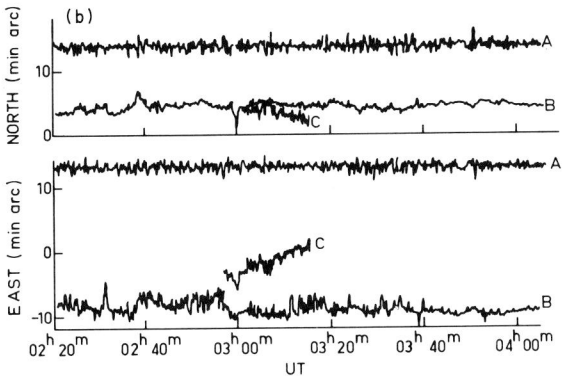

Fig. 22. An example of correction for ionospheric refraction. (a) The east–west and north–south positions (relative to the center of the solar disk) of three sources A, B, C observed on July 26, 1970 during a period of severe ionospheric disturbance. The synchronous movement of sources more than 1 $R_\odot$ apart, over distances of about 1 $R_\odot$, suggests the influence of varying ionospheric refraction. (b) The same data, corrected by assuming the position of the source A to be constant. Since the corrections are smoothed over about 1 min, this figure still shows residual jitter in the position of A.

This makes the editing procedure long and tedious, involving some rather difficult choices; on occasion valuable data have been lost through the need to make a selection before all the relevant information is available. Edited data on long-lived storm activity are always incomplete, owing to the need to keep the magnetic tape costs to a reasonable limit.

In order to solve this problem we have installed a computer which is now being interfaced to the radioheliograph output. By the time this volume is published it is expected that the computer will be processing the observations in real time and recording the data in a compressed format on magnetic tape. In this way *all* the *useful* data output by the radioheliograph will be recorded on about one 2400 ft reel of magnetic tape per day—roughly the same as the present average tape consumption for edited data. We estimate that the amount of data available for offline analysis will be increased about tenfold. Of course, this increase will compound the difficulties of analysis and processing, but the extra complexity will be amply justified by the advantages of more complete and more consistently sampled data.

The method of achieving this dramatic saving in tape is best illustrated by Fig. 23. At present data are recorded in each of two states of circular polarization on a raster of 48×60 points. In Fig. 23a a photographic reproduction of both polarizations of a typical record is shown. It is at once clear that most of the field is empty. Of course the large field is necessary because it is not possible to predict where a burst will appear; however, after the event it is perfectly reasonable to discard the empty 95% or so of the field and retain only the information from the remaining few percent. This is illustrated diagrammatically in Fig. 23b by cutting away the empty part of the field.

The validity of this procedure rests on the fact that, as previously mentioned, sources fainter than a few percent of the brightest point in the image cannot be reliably detected because of confusion with sidelobes of instrumental origin. It is not unreasonable therefore to discard all raster points fainter than this limit.

The amount of the field covered by burst sources varies greatly but Fig. 23 is fairly typical. Figure 15 is a rare example of a particularly extended source, and even here only about 20% of the field is covered.

We have seen that when there is no burst activity the radioheliograph can be used to study the thermal emission of the quiet Sun, and that it is then necessary to sum about 100 sec of data to get an adequate signal-to-noise ratio. The computer will therefore be programmed to switch automatically to an integration mode whenever the maximum brightness of the Sun falls below a level just above that of the quiet Sun. This again will represent an enormous saving in magnetic tape.

The following is the sequence of operations in the proposed compression program.

(1) Determine the brightness maximum for each polarization separately (this step is complicated by the need to discriminate against lightning, but we postpone discussion of this point until the next section).

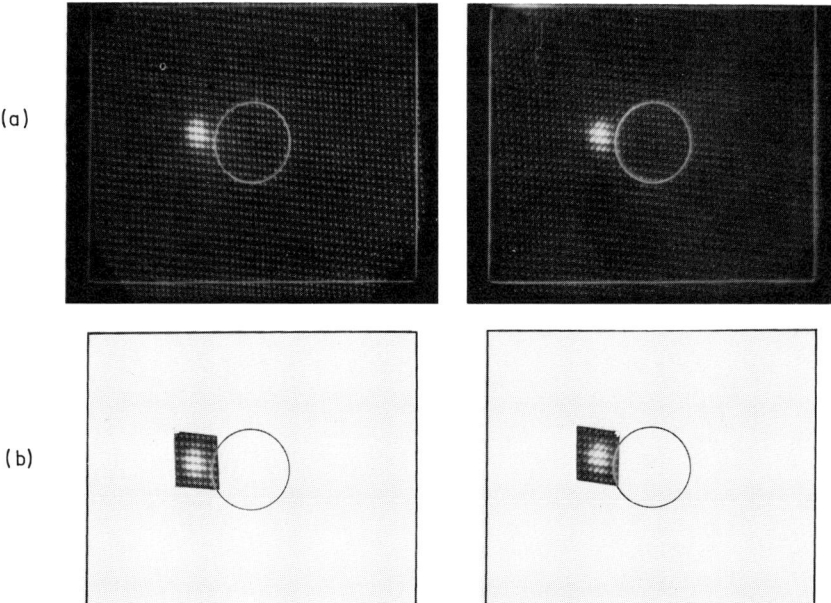

FIG. 23. (a) Two frames of a typical heliograph record. After the event we can see that most of the field is empty. (b) The "empty" parts of the same record have been cut away, leaving only the useful parts of the pictures. These figures illustrate the principle of the data compression process, performed in real time by a small on-line computer.

(2) If the maximum of each polarization is less than the integration threshold, then the record is to be integrated; jump to step 13 below. Otherwise continue for data compression with steps C3 to C7.

Compression

(C3) Divide the brightness maxima by a constant determined by the instrumental sidelobe level, to set a second, lower threshold for each polarization.

(C4) Compile a map in memory showing the positions of all points brighter than this second threshold in either polarization. To allow detailed comparisons between polarizations, data for both polarizations will be saved for any point which is above the threshold in either polarization.

(C5) Add to the map all points adjacent to a point which is already in the map; this will as a precaution extend the saved area a little beyond the bare minimum, into the "skirts" of the sources.

(C6) Add to the output buffer a data heading (date, time, etc.) followed by the data for all points in the map.

(C7) The processing of the record is now complete. In general there may be about 0.1 to 0.2 sec available at this stage for other jobs on a second-priority basis, before commencing at step 1 for the next record.

Integration

(I3) If this is the first record of a new integration the input buffer becomes the integration buffer and no processing is required.

(I4) Otherwise add the new record into the cumulative sum in the integration buffer.

(I5) Same as C7 above.

It will be appreciated that while one record is being processed the heliograph will be transferring the next record to another input buffer. The programmer must ensure that whenever the heliograph is ready to start the input of a new record a buffer is available to receive the data. This means that all the processing of a record must be completed within 1 sec, and also that the program must keep a check on the need to output a buffer to magnetic tape and initiate this output as soon as possible. The most time-consuming parts of the program have already been coded and timed and there appear to be no difficulties on this score. For a number of reasons, particularly space and speed, all the coding is being done in assembly language. The compression and integration procedures have been tested successfully on a general purpose computer, using existing full-format records.

The computer chosen for this work is a PDP-15 manufactured by Digital Equipment Corporation. This unit has 24K words of core, two-thirds of which will be taken up by four large buffers of 4K words each (one for integration, one for compression, and two for input).

2. *Interference Suppression*

An additional complication is introduced by the need to recognize and eliminate the interference produced by nearby thunderstorms. Because a lightning flash is very much briefer than the time taken to scan an image, the radio interference it produces appears as a single bright column, extending over the full height of the raster. The example shown in Fig. 24 shows that it is very easily recognized visually, and also that an occasional lightning stroke does very little harm to the usefulness of the records.

However, lightning will upset the functioning of the compression program in two ways. The first is just a nuisance—it will tend to increase the amount of data saved, all the extra data being useless lightning

FIG. 24. A radioheliograph record showing typical lightning interference. In this case the source in the center of the Sun is fainter than the bright columns on the right of the figure, which are due to brief lightning flashes. This interference is easily distinguished visually and can be recognized and eliminated from the record by the computer.

flashes. The second problem is more serious: the brightest lightning flashes may appear much brighter than any concurrent solar sources and so cause the threshold, a few percent of the brightest raster point, to be above the brightness of the genuine sources. When this happens the compression program will proceed to discard all the good data and keep the interference.

Since thunderstorms are very common at Culgoora at certain times of the year this problem is sufficiently important to require a solution. Although lightning is so easily recognized by visual inspections of radioheliograms, it is surprisingly difficult to design an efficient subroutine to perform this task. The difficulty is due to the bulk of the data to be processed and the need to finish all processing of a record in less than 1 sec. Even a very simple program loop to find the brightest point in the raster takes about 0.05 sec by the time it has been repeated 5760 times (once for each picture point). Any iterative procedure with a slightly more complicated loop would therefore be likely to use up all the available time.

The method adopted is the following. In a single pass through the data the row sums, the column sums, and the column maxima are determined. Since a lightning flash is fairly uniformly bright in all the points of a column, its only effect on the row sums will be to raise them all equally. Therefore the maximum deviation of the row sums from their mean is a good indication of the level of solar activity. After allowing a factor of 0.8, because there are 48 rows and 60 columns, and a factor of 2 or 3 to allow for the possibility of elongated sources, the maximum deviation of the column sums from their mean should not be more than about twice the maximum row deviation. Any columns which do not satisfy this criterion are simply ignored. This procedure will guarantee that the maximum is not set much higher than it should be, and will reduce the amount of tape wastage.

For quiet-Sun integration where the accumulation of occasional flashes can completely obliterate the integrated result, a rather more severe criterion will be used, with smoothing of both row and column sums. Whereas for burst work it is better to let an occasional flash through than to lose any genuine data, for quiet-Sun integration the reverse is true.

VII. Solar Radio Astronomy with the Radioheliograph

We have already briefly discussed the use of the Culgoora radioheliograph in the observation of the thermal background emission from the quiet Sun (Fig. 17). We now turn to the application of the radioheliograph to the study of solar radio bursts.

The bursts of enhanced radiation from the corona exhibit a great diversity of characteristics. The earliest method of studying their properties—and one which remains very important—is by the use of "dynamic spectrum" recordings (Fig. 25), in which the intensity of the radiation is presented (usually on photographic film) as a function of both frequency and time. Spectrum investigations led to a classification of bursts into a number of spectral types, which correspond to different exciting agencies and radiation mechanisms.

The radioheliograph now makes it possible also to study in detail the spatial structure of the sources of coronal bursts and their second-by-second evolution. In this way important advances have been made in our understanding of the mechanisms by which the burst radiation is generated; of the physical nature of the various exciting agencies; and of the plasma and magnetic-field conditions in the regions where the bursts are emitted and through which the radiation is propagated. Studies of radioheliograph observations have been published in a large number of research papers, most of which are discussed, with references, in a recent review (Wild and Smerd, **1972**). Here we shall briefly outline just a few of the results which are important in themselves and which also illustrate some applications of the methods of data analysis we have described.

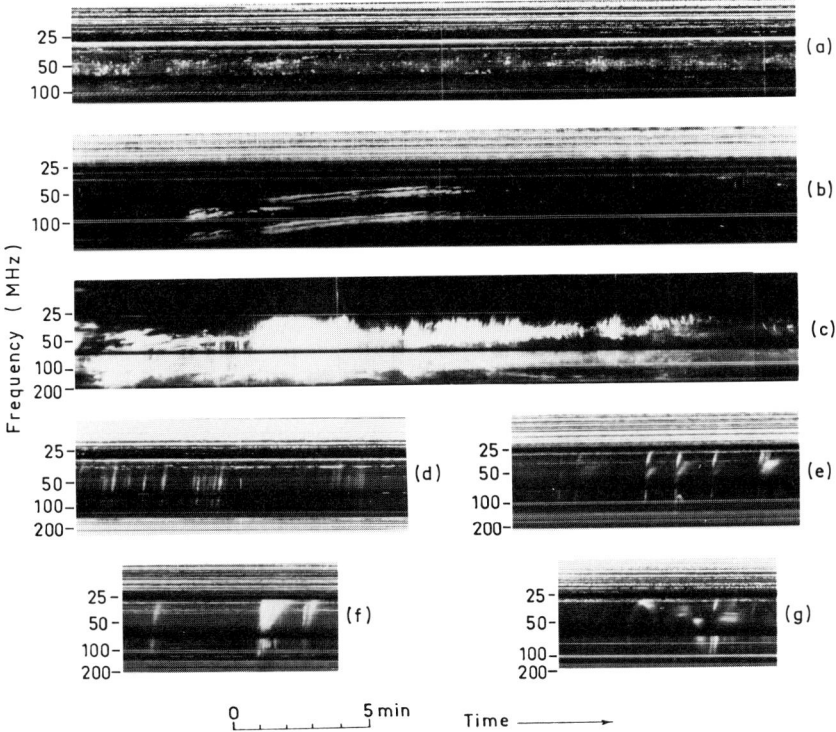

FIG. 25. Dynamic spectra of meter-wavelength solar bursts. (a) Type I storm. (b) Type II, with "split bands" and harmonic structure. (c) Type II, showing "herringbone" structure. (d) and (e) Type III groups; fundamental and second harmonic are clearly recognizable in (e). (f) Type V. (g) U-bursts (from Labrum, 1972.)

RADIO BURSTS FROM THE CORONA

1. Type I Bursts

These events (Fig. 25a) are both the commonest and the least understood of meter-wavelength bursts. It is generally accepted that type I emission is excited near the plasma level* by moderately energetic electrons trapped in the strong closed-loop magnetic fields above sunspot groups. The radiation is often strongly circularly polarized, in a sense which is found to correspond to propagation in the ordinary magnetoionic mode for the dominant magnetic field.

* The plasma level for a particular frequency is the height in the corona at which the electron plasma frequency f_0 [Eq. (2)] is equal to the radio-wave frequency.

Bipolar structure, i.e., the spatial separation of the right- and left-handed polarized components, can frequently be detected by superposing radioheliograph contour plots of type I events in the two senses of circular polarization (see Fig. 16). This effect is more clearly demonstrated by computer-generated scatter plots of successive source positions in long-lasting type I "storms." Using the scatter-plot technique, McLean and Sheridan (1972) have found (see Fig. 21) that the displacement between the two senses of polarization does not necessarily correspond to the distribution of polarity in the underlying magnetic field and may indeed show a complete reversal from the latter. This is interpreted as evidence of twisting of the magnetic-field loops in the corona below the height (0.6 $R_\odot$ at 80 MHz) of the radio sources.

2. Type III Bursts

Type III radio bursts (Fig. 25d,e) are characterized by a rapid downward drift in emitting frequency, and there is strong evidence that they are excited by streams of electrons, accelerated in solar flares to velocities of the order of $c/3$, which set up plasma waves close to the local plasma frequency as they move outward through the corona. These waves produce radio emission (fundamental or second harmonic) by scattering processes in the corona.

Radioheliograms have shown that the type III sources are invariably displaced from the sources of type I storms associated with the same active region and are located in regions of weak magnetic field (Kai, 1970). This agrees with a theoretical prediction (Weiss and Wild, 1964) that the electron streams must escape along or near magnetic neutral planes. Further confirmation is provided by the occasional observation of successive type III burst positions along the line of an optical filament; it is known that filaments usually delineate magnetic neutral planes.

In some variants of type III bursts, the spectra give evidence that the electrons are injected into closed magnetic loops, to be lost at the end of the loop ("U"-bursts) or briefly trapped (type V). This interpretation led to the prediction of position shifts during the evolution of type V bursts and U bursts; this has now been confirmed by radioheliograph observations.

In some cases type III spectra show both fundamental and second harmonic components. The plasma-wave hypothesis leads us to expect that at a given frequency the fundamental is emitted at the plasma level and the harmonic further out at the half-frequency plasma level. However, radioheliograms show that the observed source positions are nearly always very close together (Stewart, 1972). This apparent paradox has been explained in terms of the propagation of the radiation outward through the corona. An early interpretation (Smerd et al., 1962) attributed the observed structure to the predominantly backward emission of the harmonic radiation from its source region. However, recent ray-tracing computations (Steinberg et al., 1971; Riddle, 1972a,b, 1974; Leblanc, 1973) which take into account both scattering and refraction in the corona, show that these factors may well explain the apparent coincidence of the fundamental and harmonic sources without any assumption of anisotropic emission.

3. Type II Bursts

These spectacular radio events (Fig. 25b and c) are excited by coronal shock waves spreading from a flare explosion. The shock waves can occasionally be detected optically (Moreton, 1964; Ramsey and Smith, 1966); the slow frequency drift of the radio bursts is consistent with the predicted outward velocity (~ 1000 km/sec) of the shock waves. The plasma waves which produce the radio emission are set up by electrons accelerated in the shock front; in some cases, such as Fig. 25c, fast-drift bursts corresponding to individual pulses of electrons give the spectrum of the type II burst a "herring-bone" appearance.

Radioheliograms show that type II sources are typically of large dimensions ($\sim \frac{1}{2} R_\odot$ at 80 MHz) and have rapid fluctuations in their brightness distribution. In several cases three or four sources have appeared around a wide arc centered on the site of the flare, as shown by sources A, C, and D in Fig. 26 (Smerd, 1970) (one stage of this spectacular event is also shown in Fig. 15). Examples such as this show that the type II

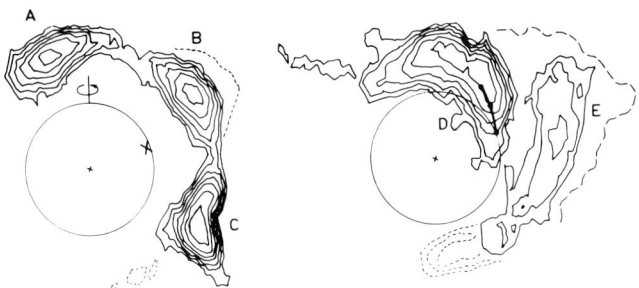

FIG. 26. Great limb event of March 30, 1969. Circumstantial evidence indicates that the flare (a proton event) was located *behind* the limb at position X starting at $\sim 02^h 45^m$ UT. The 80 MHz heliograms shown were taken at $02^h 50^m$ (*left*) and $03^h 03^m$-06^m. Type II bursts occurred in sources A, C, and D. Source B is a continuum source (from Smerd, 1970).

emission is restricted to particular sections of the spreading shock front (Uchida, 1968, 1970, 1973; Uchida et al., 1973).

An interesting feature of many type II spectra is the presence of two parallel emission bands in both the fundamental and harmonic components (Fig. 25b). Radioheliograph observations show that the upper and lower bands correspond to two successive bursts from separate positions at a particular frequency. Thus any valid model of the emission process must explain both the split-frequency spectrum and the fact that the two spectral components are emitted in succession and from different source positions. McLean (1967) has suggested that the "split-band" structure corresponds to preferred emission from two points where the shock front is most nearly tangential to the levels of constant plasma density (Fig. 27a). It is indeed found that a simulation of this situation (with plausible assumptions for the values of the various physical parameters) yields a computer-generated "spectrum" (Fig. 27b) which reproduces the main spectral features of recorded split-band bursts. Alternatively, Smerd et al. (1973) have recently suggested that the two bands

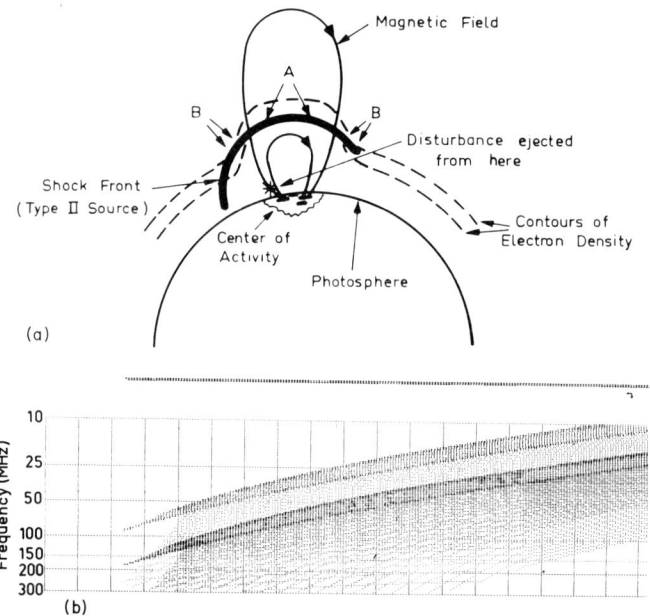

FIG. 27. Split-band type II bursts. (a) A suggested source model. The two sources of type II emission are in the regions where the expanding shock front is most nearly tangential to the contours of equal coronal electron density. (b) A computer-simulated type II burst, based on this model (cf. Fig. 25b). The horizontal divisions correspond to time intervals of 1 m (from McLean, 1967).

may represent emission from two regions, in front of and behind the shock front, where the plasma densities are different.

4. Moving Type IV Bursts

In the late stage of the radio "outburst" which follows a large solar flare, the spectrum often shows a wideband "continuum" radiation. Early fan-beam observations indicated that the source of this component is located high in the corona and often shows an outward movement during its lifetime (a few minutes to an hour or so). This emission, which is known as a type IV burst, is attributed to synchrotron radiation from a cloud of electrons with relativistic energies, spiralling around magnetic field lines. It is in the elucidation of these often complicated type IV events that the pencil-beam radioheliograph has made perhaps its most important contribution to solar physics.

Combined spectrograph and radioheliograph observations have shown that one variety of type IV burst is associated with the same shock front which excited the preceding type II bursts. Thus Kai (1969) showed that in one instance the whole event can be explained in terms of a spreading shock front (Fig. 28); this first produced a type II burst observed at the 80 MHz plasma level (height $\sim 0.6 \ R_\odot$) and later the type IV burst at about 1.0 $R_\odot$. The fact that the synchrotron radiation is not observed until the front is well beyond the plasma level confirms a theoretical prediction (Zheleznyakov and Trakhtengerts, 1965) that the low-frequency part of the synchrotron spectrum (below about twice the local plasma frequency) will be effectively suppressed by the medium.

Another class of type IV burst, such as the example in Fig. 29a (Wild 1969), appears to be linked with the "piston" of coronal gas expanding behind the shock front. In this case the emission is associated with an arch of magnetic field, expanding with the outward-moving gas. Electrons are accelerated into this arch, perhaps by the rising shock front. There they emit synchrotron radiation at the top of the arch while radiation, circularly polarized in the two opposite senses, is emitted from two sources near the plasma level at the feet of the arch.

In a third class of type IV burst the radiation can be attributed to large magnetized clouds of solar gas moving outward from the Sun. The radioheliograph has for the first time enabled us to observe these "far-moving" type IV events with sufficient angular resolution to reveal their complex spatial structure. It has been found that the sources often consist of two or more components with differing degrees and senses of circular polarization. An example is shown in Fig. 29b (Riddle, 1970) in which the isolated moving radio source was tracked to about 4 $R_\odot$ from the Sun.

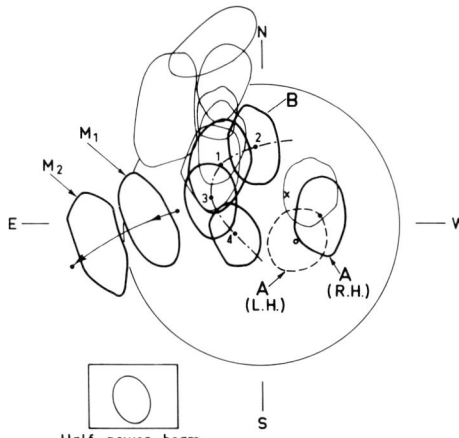

Fig. 28. Summary of 80 MHz source configurations during a solar outburst on September 4, 1968. Sources B (1, 2, 3, 4) correspond to type II bursts and are considered to be located at points on the shock front expanding from X, the site of the flare. M_1 and M_2 are successive positions of a moving type IV burst (path indicated by arrows); the direction of motion and time of occurrence indicate an association with the same shock front. A circularly polarized stationary type IV burst (A) and a number of type III bursts (thin contours) were also observed (from Kai, 1969).

This radio event followed the optical observation (at the Mount Haleakala observatory of the University of Hawaii) of a spectacular ejection of material to a height of about 1 $R_\odot$; the time lag of about an hour between the optical and radio events is difficult to explain. In other events of this

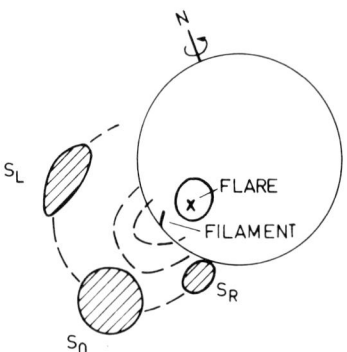

Fig. 29a. A moving type IV burst of the "expanding magnetic arch" variety. On the radioheliograms 80 MHz emission appeared first around the inner arches; later the emission was from the three discrete sources S_0 (unpolarized), S_L (left-handed), and S_R (right-handed). The later storm phase occurred in the unshaded source near the flare (from Wild, 1970).

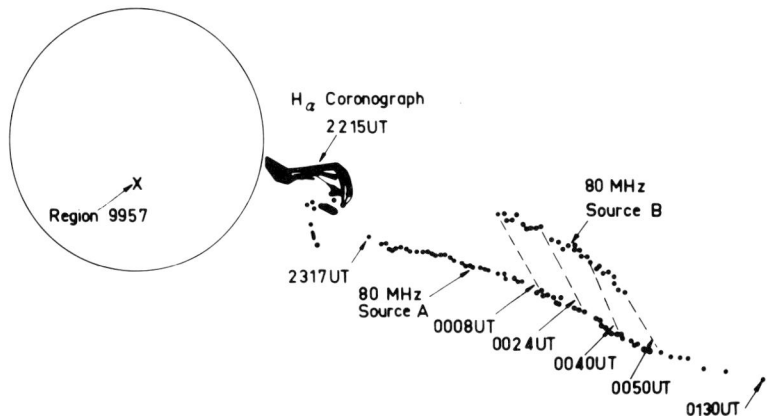

FIG. 29b. The path of a "far-moving" type IV radio burst observed at 80 MHz. The associated optical eruption is also shown (from Riddle, 1970).

class (e.g., Sheridan, 1970) more complicated distributions of circularly polarized sources have been observed. In many cases the degree of circular polarization of the individual components increases toward the end of the event, reaching values as high as 60–90%.

Now, synchrotron radiation from electrons of very high energy is linearly polarized; the occurrence of strong circular polarization can therefore be taken to indicate that the type IV electrons have only moderately relativistic energy, in the range about 0.1–1 MeV. The observed high circular polarization in the later stages of far-moving type IV events has been accounted for by a theory proposed by Dulk (1970) and extended by Schmahl (1972) and Dulk (1973) in which the radiation is taken to be from electrons in an orderly and gradually expanding magnetic field.

VIII. Future Developments in Radioheliography

It is clear from this discussion that the solar radio astronomer would like future radioheliographs to combine the highest resolution in both time and angular position with observations over the widest possible range of frequencies. At the same time the output data, the volume of which will become larger and larger as these ideals are approached, must be suitable for presentation in readily assimilable forms for astronomical studies. For signal processing it seems likely that scanning systems, which are inherently slow and inefficient, will be superseded by one of the direct-imaging techniques discussed in Section III,C. The use of digital recording and data-handling methods appears essential for the necessary flexibility in methods of analysis. With ever-increasing data rates, preliminary data reduction by on-line computers will become more and more necessary to avoid the problem of storing, and subsequently processing, unmanageable quantities of raw data.

ACKNOWLEDGMENTS

The Culgoora radioheliograph, which is basic to much of the work described in this article, was developed by a large group of our colleagues in the Division of Radiophysics. We would like particularly to acknowledge the major contributions of Mr. M. Beard, who designed and supervised the construction of the digital control and processing sections; Dr. K. V. Sheridan, who was responsible for the receiver system and the coordination of the whole construction project; and Messrs. A. F. Young and W. J. Payten, who have also made important contributions to the development of the digital systems.

We thank Dr. S. F. Smerd and Dr. J. A. Roberts for many valuable discussions and suggestions during the writing of this chapter.

The construction of the radioheliograph was made possible by generous grants, totaling $630,000, from the Ford Foundation.

REFERENCES

BLUM, E. J. (1961). *Ann. Astrophys.* **24**, 359.
BRACEWELL, R. N., and SWARUP, G. (1961). *IRE Trans. Antennas Propagat.* **9**, 22.
BRIGGS, B. H., and HOLMES, N. (1973). *Nature (London), Phys. Sci.* **243**, 111.
CARTER, A. W. L., and WILD, J. P. (1964). *Proc. Roy. Soc., Ser. A* **282**, 252.
CHIUDERI, C., CHIUDERI DRAGO, F., and NOCI, G. (1971). *Solar Phys.* **17**, 367.
CHRISTIANSEN, W. N. (1953). *Nature (London)* **171**, 831.
CHRISTIANSEN, W. N., and MATHEWSON, D. S. (1958). *Proc. IRE* **46**, 127.
CHRISTIANSEN, W. N., and WARBURTON, J. A. (1953). *Aust. J. Phys.* **6**, 262.
CHRISTIANSEN, W. N., and WARBURTON, J. A. (1955). *Aust. J. Phys.* **8**, 474.
COVINGTON, A. E., and BROTEN, N. W. (1957). *IRE Trans. Antennas Propagat.* **5**, 247.
DULK, G. A. (1970). *Proc. Astron. Soc. Aust.* **1**, 372.
DULK, G. A. (1973). *Solar Phys.* **32**, 491.
ERICKSON, W. C. (1973). *Proc. IEEE* **61**, 1276.
GABOR, D. (1969). *Opt. Acta* **16**, 519.
KAI, K. (1969). *Solar Phys.* **10**, 460.
KAI, K. (1970). *Solar Phys.* **11**, 456.
KUNDU, M. R., ERICKSON, W. C., JACKSON, P. D., and FAINBERG, J. (1970). *Solar Phys.* **14**, 394.
LABRUM, N. R. (1971). *Aust. J. Phys.* **24**, 193.
LABRUM, N. R. (1972). *In* "Solar Activity Observations and Predictions" (P. S. McIntosh and M. Dryer, eds.), p. 93. MIT Press, Cambridge, Massachusetts.
LAMBERT, L. B., ARM, M., and AIMETTE, A. (1965). *In* "Optical and Electro-Optical Information Processing" (J. T. Tippett *et al.*, eds.), Chapter 38, p. 715. MIT Press, Cambridge, Massachusetts.

LEBLANC, Y. (1973). *Astrophys. Lett.* **14**, 41.
MCLEAN, D. J. (1967). *Proc. Astron. Soc. Aust.* **1**, 47.
MCLEAN, D. J. (1973). *Proc. Astron. Soc. Aust.* **2**, 222.
MCLEAN, D. J., and SHERIDAN, K V. (1972). *Solar Phys.* **26**, 176.
MCLEAN, D. J., and WILD, J. P. (1961). *Aust. J. Phys.* **14**, 489.
MCLEAN, D. J., LAMBERT, L. B., ARM, M., and STARK, H. (1967). *Proc. Inst. Radio Electron. Eng. Aust.* **28**, 375.
MILLS, B. Y., and LITTLE, A. G. (1953). *Aust. J. Phys.* **6**, 272.
MORETON, G. E. (1964). *Astron. J.* **69**, 145.
O'BRIEN, P. A. (1953). *Mon. Not. Roy. Astron. Soc.* **113**, 597.
RAMSEY, H. E., and SMITH, S. F. (1966). *Astron. J.* **71**, 197.
RIDDLE, A. C. (1970). *Solar Phys.* **13**, 448.
RIDDLE, A. C. (1972a). *Proc. Astron. Soc. Aust.* **2**, 98.
RIDDLE, A. C. (1972b). *Proc. Astron. Soc. Aust.* **2**, 148.
RIDDLE, A. C. (1974). *Solar Phys.* **35**, 491.
SCHMAHL, E. J. (1972). *Proc. Astron. Soc. Aust.* **2**, 95.
SHERIDAN, K. V. (1970). *Proc. Astron. Soc. Aust.* **1**, 376.
SHERIDAN, K. V., LABRUM, N. R., and PAYTEN, W. J. (1973). *Proc. IEEE* **61**, 1312.
SMERD, S. F. (1950). *Aust. J. Sci. Res., Ser. A* **3**, 34.
SMERD, S. F. (1970). *Proc. Astron. Soc. Aust.* **1**, 305.
SMERD, S. F., WILD, J. P., and SHERIDAN, K. V. (1962). *Aust. J. Phys.* **15**, 180.
SMERD, S. F., SHERIDAN, K. V., and STEWART, R. T. (1973). *In* "Coronal Disturbances" (G. Newkirk, ed.), Proc. Int. Astron. Union Symp. No. 57. Reidel Publ., Dordrecht, Netherlands (in press).
STEINBERG, J. L., AUBIER-GIRAUD, M., LEBLANC, Y., and BOISCHOT, A. (1971). *Astron. Astrophys.* **10**, 362.
STEWART, R. T. (1972). *Proc. Astron. Soc. Aust.* **2**, 100.
TAKAKURA, T., TSUCHIYA, A., MORIMOTO, M., and KAI, K. (1967). *Proc. Astron. Soc. Aust.* **1**, 56.
TANAKA, H., and KAKINUMA, T. (1954). *Proc. Nagoya Univ. Inst. Atmos.* **2**, 53.
TANAKA, H., and KAKINUMA, T., ENOMÉ, S., CHIKAYOSHI, T., TSUKIJI, Y., and KOBAYASHI, S. (1969). *Proc. Nagoya Univ. Inst. Atmos.* **16**, 113.
UCHIDA, Y. (1968). *Solar Phys.* **4**, 30.
UCHIDA, Y. (1970). *Publ. Astron. Soc. Jap.* **22**, 341.
UCHIDA, Y. (1973). *In* "High Energy Phenomena on the Sun" (R. Ramaty and R. G. Stone, eds.), Nat. Aeron. Space Admin., Washington, D.C. NASA-GSFC Preprint X-693-73-193, p. 577.
UCHIDA, Y., ALTSCHULER, M. D., and NEWKIRK, G. (1973). *Solar Phys.* **28**, 495.
VINOKUR, M. (1968). *Ann. Astrophys.* **31**, 457.
WEISS, A. A., and WILD, J. P. (1964). *Aust. J. Phygs.* **17**, 282.
WILD, J. P. (1961). *Proc. Roy. Soc., Ser. A* **262**, 84.
WILD, J. P. (1965). *Proc. Roy. Soc., Ser. A* **286**, 499.
WILD, J. P. (1967). *Proc. Inst. Radio Electron. Eng. Aust.* **28**, 277–384.
WILD, J. P. (1970). *Proc. Astron. Soc. Aust.* **1**, 365.
WILD, J. P., and SMERD, S. F. (1972). *Annu. Rev. Astron. Astrophys.* **10**, 159.
ZHELEZNYAKOV, V. V., and TRAKHTENGERTS, V. YU. (1965). *Astron. Zh.* **42**, 1005; *Sov. Astron.—AJ* **9**, 775 (1966).

Pulsar Signal Processing

TIMOTHY H. HANKINS[*] AND BARNEY J. RICKETT

DEPARTMENT OF APPLIED PHYSICS AND INFORMATION SCIENCE
UNIVERSITY OF CALIFORNIA AT SAN DIEGO
LA JOLLA, CALIFORNIA

I. Introduction	56
II. Pulsar Searches	57
A. The Fast Folding Algorithm (FFA) Method	58
B. The Fast Fourier Transform (FFT) Method	60
C. Comparison of Digital Period Search Techniques	61
D. Digital Dispersion Search Techniques	62
III. Dispersion	64
A. Transfer Function of the Interstellar Medium	64
B. The Effect of Dispersion on Time Resolution	67
C. Dispersion Removal	69
D. Measurement of Dispersion	77
IV. Sampling, Resolution, and Average Profiles	81
A. Instrumental Considerations	81
B. Average Profiles	83
V. Polarization	84
A. Polarimeter Calibration	85
B. Polarization Observations and Displays	86
VI. Intensity Variations with Time	87
A. Intensity Variations and Their Statistics	87
B. Correlation Functions for Variations within a Pulse Period	93
C. Pulse-to-Pulse Correlation	96
D. Pulse Energy Fluctuations	97
E. Fluctuation Spectra versus Longitude	98
F. Intensity Distribution Functions	102
G. Estimation Error in Mean Profiles	103
VII. Intensity Variations with Frequency	104
A. Frequency Dependence of Pulsar Radiation	104
B. Frequency and Time Resolution	106
C. Narrowband Spectral Variations	107
VIII. Interstellar Scattering and Scintillation	109
A. Fluctuations Over Time and Frequency	109
B. Pulse Broadening	112
C. Multiple-Site Observations	116

[*] Present address: Arecibo Observatory, Box 995, Arecibo, Puerto Rico 00612.

IX. Timing Measurements 118
    A. Time-of-Arrival Observations 118
    B. Errors in Arrival Time 119
    C. Barycentric Arrival Times 121
    D. Model Fitting 122
    References 126

I. Introduction

IT IS INCONGRUOUS THAT Jocelyn Bell would not have discovered pulsars (Hewish et al., 1968) if she had used a computer for her survey of scintillating radio sources, but that ever since their discovery almost all pulsar observations have relied heavily on computers for data processing. This article is devoted to the various techniques which have been used in searching for and processing the pulsar signals. The methods are mostly routine statistical operations on large amounts of data. The complexity comes more from the range of variability contained in the pulsar signals; there are still no adequate physical explanations for most of these variations.

The most striking feature of pulsar signals is their extremely stable periodicity, departures from a constant period being measurable only by reference to atomic time standards over long intervals. The periods of the more than 100 known pulsars lie in the range from 0.033 to 3.7 sec, and the duration of each pulse is typically 2–5% of the period. In contrast to the stability of the period, the radio pulse intensity varies enormously, with time scales from months down to microseconds.

Sampling pulsar signals at a rate fast enough to resolve microsecond time structure presents an awesome data handling and display problem. Even if the signal is sampled only "on pulse" during about 5% of its period, as many as 10^4 independent data values can be obtained for each pulse, and if the full polarization state is desired, then the signal from an orthogonally polarized antenna must also be recorded simultaneously. Although high time resolution studies are essential for determining details of the pulsar emission, it is clearly impractical to study individual pulses with microsecond resolution on a routine basis. Furthermore, many pulsars are so weak that they can only be detected by averaging the pulses synchronously with the period P_1 to form the average pulse profile.

The various averaging techniques, which have been developed to compress the huge amounts of data available from a single pulsar, are the subject of this article. We will concentrate mainly on the data processing methods, displaying the results and to a lesser extent on their physical implications. Searching for the periodic and/or dispersed pulses is discussed in Section II. Interstellar dispersion, its influence on time resolution, methods for measuring it, and removing its effects are discussed in Section III. Section IV outlines resolution and sampling problems and the computation of the average pulse profile. In general we assume that the signals have been adequately sampled to avoid loss of information from aliasing. The methods for polarization observations are given in Section V. Subsequent sections (VI to IX) discuss intensity variations over time scales from microseconds to millions of years and over radio frequency ranges from 40 MHz to 10 GHz. We give greatest detail in areas of our own experience (namely dispersion removal and intensity variations) and refer more to the published literature in other areas.

II. Pulsar Searches

Many methods have been used to search for pulsars—from visual examination of intensity chart records to multidimensional computer searches. The first pulsar (PSR 1919+21) was discovered by noting spikes on a chart record (Hewish et al., 1968), and a substantial number of other pulsars were found this way (Vaughan and Large, 1969). The weaker and more sporadic pulsars, however, either cannot be seen at all on ordinary chart records, or cannot easily be distinguished from impulsive interference. Although the number of dimensions to search is large (two spatial coordinates, radio frequency, period, pulse width, pulse phase, and dispersion), pulsar signals have such distinct signatures in some of these dimensions that their characteristics can be used to discriminate even very weak pulsars from other sources of radio emission. Search methods which have been used will be discussed and compared briefly; they have been described in more detail in the literature.

Searches in spatial coordinates have been dictated partly by antenna pointing restrictions and partly by telescope time opportunities. After only a dozen pulsars had been found there appeared to be a clustering of pulsars in the galactic plane (Wielebinski et al., 1969). Since pulsars had been determined to be galactic objects, considerable effort has been spent in searching the galactic plane for more pulsars, with notable success (Hulse and Taylor, 1974).

Although the first four pulsars were discovered at 81.5 MHz, most pulsars have been discovered at frequencies around 400 MHz. Most pulsars are strongest between 100 and 400 MHz (Comella, 1972), but interstellar scintillation causes random, slowly varying, deep modulation of the frequency spectrum as described in Section VIII. In addition, pulsars show intrinsic intensity variations on scales of seconds to months.

There exists the possibility of missing detection of a pulsar while scanning its position in the sky at a time coincident with an intrinsic or interstellar scintillation fade. Frequency and time diversity can be used to overcome this problem.

The most striking feature of pulsar radiation is the constant period of the intensity pulsations. If one assumes that the pulses are rectangular and of uniform amplitude and width, then optimum detection in the presence of Gaussian noise is obtained by convolution of the received detected signal with the matched filter whose impulse response is described by rectangular functions spaced at the assumed period. When the period is known *a priori*, this is sometimes called "rail filtering," and it is discussed in Section IV. In its simplest form, a series of detector output samples taken at intervals Δt is divided into strings of length equal to the assumed period P_1. Then these strings are aligned to the nearest sample and summed to give an average pulse profile, which is scanned for significant pulse peaks. If there are N samples in the data set, N additions are required for each assumed period P_1, and the period must be searched in increments of P_1/N. For long data sets this method is very cumbersome since the number of operations increases as N^2.

Two approaches to a more efficient period search have been developed: the fast folding algorithm (FFA) (Staelin, 1969) and the fast Fourier transform (FFT) (Lovelace et al., 1969) methods. These have been reviewed and compared by Burns and Clark (1969) and are summarized below.

A. The Fast Folding Algorithm (FFA) Method

Suppose we want to search a long data series of N samples from a receiver detector output for a pulsed signal with a period between $P\Delta t$ and $(P + 1)\,\Delta t$, where P is an integer and Δt is the sampling interval. If we assume that the pulse width is less than or equal to Δt, then for a data series containing a large number M of pulses such that $MP \approx N$, the best period resolution that can be obtained is $\Delta t/M$, and there are M distinguishable periods in the range $P\Delta t$ to $(P + 1)\,\Delta t$. In a period search, then, the data samples nearest the appropriate intervals are added together, and the sums are compared with a threshold determined from the off-pulse noise level. For each of the M periods between P and $P + 1$ there are P possible arrival times for a total of $PM = N$ sums which must be tested for significance. A pulsed signal of uniform amplitude will appear in one of the sums and will grow as $\sqrt{M}$ relative to off-pulse sums. The number of additions required to produce the N sums is MN. These sums can be combined to produce sums for periods of $\frac{1}{2}$, $\frac{1}{4}$,

$\frac{1}{8}$, ..., of the period of the original pulse sums; therefore, the sums need be calculated only over some octave range of periods $P_0/2 \leq P \leq P_0$, permitting a period search from $2\Delta t$ to $P_0\Delta t$. By adding together adjacent sums the sensitivity can be increased for pulses wider than Δt. For a pulse width of $2\Delta t$, then, two adjacent sums are added; for a pulse width of $4\Delta t$, adjacent $2\Delta t$ sums are added, and so on until the maximum pulse width of $P\Delta t/2$ is reached. The total number of operations required, including all the lower octaves and longer pulse widths, is $N^2/2$.

By using the fast folding algorithm developed by Staelin (1969), some of the redundant additions in the period search can be eliminated. This substantially reduces the number of operations required in the rail filter process. It is basically a binary tree algorithm where samples are added in pairs, then pairs are added to form fours, then fours are added to form eights, etc. The following description of the FFA is taken from Lovelace et al. (1969).

We would like to search for a pulsar with period between $P \Delta t$ and $(P + 1) \Delta t$ in a record of length greater than $N = 2^k$ intensity samples; $I(i)$ ($i = 1, 2, \ldots, N$; $\langle I \rangle = 0$) recorded at intervals Δt seconds, where P is in the octave

$$P_0/2 \leq P \leq P_0 = 2^l,$$

resulting in

$$M_0 = N/P_0 = 2^{k-l}$$

distinguishable periods. There are $k - l - 1$ stages in the process; each stage corresponds to a level $1 \leq j \leq k - l - 1$ in the binary tree algorithm.

For the first stage, form n pairs of sums

$$A_1(n, m) = I[(2n - 2)P + m] + I[(2n - 1)P + m + 0]$$
$$A_2(n, m) = I[(2n - 2)P + m] + I[(2n - 1)P + m + 1],$$

where the index $m = 1, 2, \ldots, P + M_0 - 2$ represents the pulse arrival time or phase during an interval of one pulse period, and $n = 1, 2, \ldots, M_0/2$. The A_1 terms include the sums for all the periods in the range $P \Delta t$ to $(P + \frac{1}{2}) \Delta t$, and the A_2 terms include sums for all the periods $(P + \frac{1}{2}) \Delta t$ to $(P + 1) \Delta t$.

In the second stage of addition, four kinds of sums are formed:

$$B_1(n, m) = A_1(2n - 1, m) + A_1(2n, m + 0)$$
$$B_2(n, m) = A_1(2n - 1, m) + A_1(2n, m + 1)$$
$$B_3(n, m) = A_2(2n - 1, m) + A_2(2n, m + 1)$$
$$B_4(n, m) = A_2(2n - 1, m) + A_2(2n, m + 2),$$

where $m = 1, 2, \ldots, P + M_0 - 4$, and $n = 1, 2, \ldots, M_0/4$. The four sums $B_1, \ldots, B_4$ include the periods in the ranges P to $P + \frac{1}{4}, \ldots, P + \frac{3}{4}$ to $P + 1$, respectively.

Similarly, there are 2^j kinds of sums in the jth stage with phase $m = 1, 2, \ldots, P + M_0 - 2^j$, and $n = 1, 2, \ldots, 2^{k-l-j}$. In the final stage there are M_0 sums, each having P different phases, and each with index $n = 1$. These M_0 sums correspond to the M_0 periods within the range P to $P + 1$.

Then for the fundamental octave, and the lower octave and longer pulse widths which can be derived from the fundamental, the total number of steps of addition required is

$$\frac{N^2}{2}\left[\frac{\log_2 (M_0/2)}{P_0} + \frac{3 \log_2 (8M_0 P_0^{1/2})}{4M_0}\right], \qquad (1)$$

which should be compared with the number of steps required using a direct calculation, $N^2/2$.

Inserting the parameters used at the Arecibo Observatory by Lovelace, $N = 4096$, $P_0 = 64$, or $k = 12$ and $l = 6$. Then in the octave range of periods from 32 to 64 sample points, $M_0 = N/P_0 = 64$ pulses are added; in the first lower octave, from 16 to 32 sample points from 128 pulses are added, etc. For a sample rate of 50 sec^{-1}, 4096 samples correspond to 80 sec of data with periods from 0.03 to 1.28 sec. Substituting these numbers into Eq. (1), the total number of steps required is about 2×10^6, and there are about $\frac{9}{4}NP_0 \approx 0.6 \times 10^6$ sums to be inspected for significance. If the FFA were not used, then $N^2/2 \approx 8 \times 10^6$ operations would be required. Using a computer with 2.5-μsec add time, Lovelace's program could perform the computations, exclusive of data input and output in about 75% of the time required to sample the data.

B. The Fast Fourier Transform (FFT) Method

The fast Fourier transform search method relies on the periodic nature of the pulsar being revealed in the power spectrum of the detected signal. The detected outputs from one or more receiver channels are smoothed and sampled at intervals Δt. Then the Fourier transform of a long sequence of N intensity samples from each channel is computed via the

FFT. By summing the squares of the real and imaginary Fourier coefficients, $N/2$ power spectrum coefficients are obtained, which represent the fluctuating energy in each frequency interval from 0.0 Hz to the Nyquist limit $(2\Delta t)^{-1}$. The power coefficients from several receiver channels can be added together, since the phase differences caused by dispersion delay between receiver channels are lost in the squaring operation. The square root of each of the power amplitudes is then computed to give voltage amplitude.

The spectrum coefficients could be examined directly for presence of a pulse, but the search sensitivity can be improved by summing up to $1/2D$ harmonics of the fundamental period P, where $D = W/P$ is referred to as the duty cycle, and W is the pulse width. As an example, Huguenin and Taylor (Burns and Clark, 1969) used $\Delta t = 20$ msec, $N = 2048$ and limited the period search to 2000 intervals in the range 8 to 128 points per cycle (0.16 to 2.56 sec). The block length $N\Delta t$ corresponds to about 41 sec, or roughly the time for a source to move through the fixed antenna beam they used. The harmonics are added together using a binary tree method similar to the FFA. For periods smaller than 0.64 sec, where D harmonics are not available, a smaller number of harmonics was used.

For each of the 2000 frequency intervals, a confidence level was obtained by computing the observed deviation from the mean as a fraction of the rms, where the mean and rms levels are determined from low-, medium-, or high-frequency intervals of the spectrum, depending on the frequency of the test interval. Periods which exceed the confidence limits by some arbitrary level are then selected for a rail filter search using the FFA.

C. Comparison of Digital Period Search Techniques

Burns and Clark (1969) show that, in the case where a pulsar signal is a uniform rectangular pulse train containing N pulses, sampled at intervals Δt equal to the pulse width, the signal-to-noise (S/N) ratio improvement provided by the rail filter or cross-correlation technique is

$$(S/N)_{\text{out}} = \sqrt{N}\, (S/N)_{\text{in}}.$$

Using the FFT technique and adding together the amplitudes (sum of squares of the real and imaginary parts) of L harmonics, the signal-to-noise ratio for the idealized rectangular pulse train is

$$(S/N)_{\text{out}} = \beta \left(\frac{LND}{2 + [1/ND(S/N)^2_{\text{in}}]}\right)^{1/2} (S/N)_{\text{in}},$$

where β is a function of both pulse shape and the number of harmonics averaged: $\beta = 0.77$ for rectangular pulses with spectral components summed through the $1/2D$ harmonics (Burns and Clark, 1969).

The FFA technique requires more time than the FFT technique when searching a wide range of periods, but it has better sensitivity. As the range of periods to be searched is reduced, however, the FFA takes less and less time, becoming faster than the FFT when the period range is less than an octave. Burns and Clark also show that, in the case of a pulse train whose amplitude is highly variable, these methods may be no more sensitive than single pulse searches where dispersion is sought as the identifying characteristic.

D. Digital Dispersion Search Techniques

Using a computer-assisted single pulse search technique, Davies and Large (1970) discovered three pulsars at Jodrell Bank. They computed an exponentially weighted running mean and variance of the detected outputs from receivers tuned to 406 and 410 MHz. When the (signal-mean)/rms exceeded a preset threshold of 2.6 in the 410-MHz channel, then the largest signal from the subsequent 55 samples from the lower frequency channel was selected. If it also exceeded the threshold, then an "event" was recorded by plotting a V-shaped mark on a plotter, with the length of the legs of the V proportional to the excess signal intensity over the threshold level. The V was positioned in dispersion vs. clock time coordinates, the dispersion estimate having been obtained from the delay time between the signal peaks which exceeded the threshold. The sampling interval was chosen so that the search was sensitive to dispersion measures in the range $0 \leq DM \leq 1100$ pc cm^{-3} in steps of 20 pc cm^{-3}. If several V's were plotted successively at the same dispersion, then the presence of a pulsar was suspected and the coordinates were noted for a subsequent period search.

Although their technique is in principle no more sensitive than visual examination of a chart record of the same data, the preselection of "events" by the computer greatly reduces the tedium of the search and enables the theoretical sensitivity limit to be more closely approached.

Another highly successful digital search technique based on the dispersed nature of pulsar signals has been developed by Taylor (1974a). He uses a receiver with 32 adjacent frequency channels. A dispersed pulse will appear at the output of each channel delayed in time from the adjacent higher frequency channel according to Eq. (11). The detected output from each channel is sampled at intervals Δt, and then the samples are combined with appropriate delays to match dispersions appropriate for $0 \cdot \Delta t$ sec delay across all 32 channels, $1 \cdot \Delta t$ sec delay, and so on up to $31 \cdot \Delta t$ sec delay. The samples are added together using a binary tree algorithm with delays similar to the FFA. A block diagram of the 8-channel version is given in Fig. 1. The detected filter bank outputs are sampled at intervals of Δt and stored in an array represented by the boxes on the left. The large numbers indicate the channel number in order of increasing frequency from which the sample came, and the subscript indicates the number of delays which the sample has suffered before addition in the rectangles. The delays are represented by circled τ's in the diagram. They are "holding" locations in the program which have the effect of delaying the sample by Δt sec. From the diagram it is seen that the output of the lower right box would be $I_1(t) + I_2(t - \Delta t) + I_3(t - 2\Delta t) + \cdots + I_8(t - 7\Delta t)$, where $I_i(t)$ is the signal sample at time t in receiver channel i.

It is obvious from Fig. 1 that the algorithm is modular and can be expanded for N equal to any integer power of 2. For example, the diagram shown could be expanded to $N = 16$ by replicating the structure for $N = 8$, then adding a final set of delays and additions. Note that the number of operations required increases as $N \log_2 N$.

Taylor's program requires $(3N + 5N \log_2 N + N^2)/4$ data storage locations plus about $4N \log_2 N$ locations for instructions. Then in a computer whose instruction execution time is about 3 μsec, the $N = 32$ algorithm can be executed in about 2 msec, which is considerably shorter than the typical sampling interval $\Delta t = 20$ to 50 msec. To expand the range of dispersions which may be examined, Taylor can run a second copy of the algorithm in the computer at a different rate, such as $2\Delta t$, $4\Delta t$, etc.

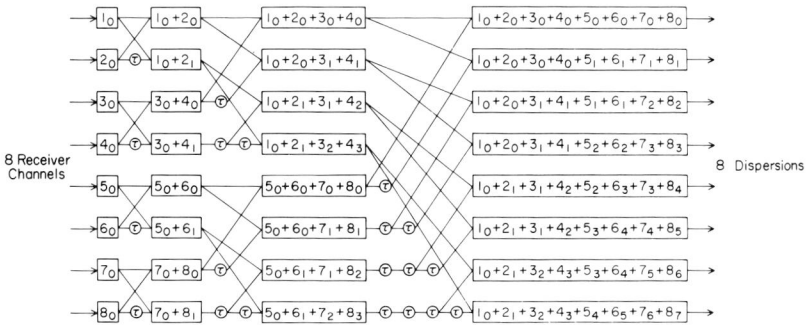

FIG. 1. Binary tree algorithm for dispersion search for an 8-channel system. The large numbers represent the input channel number. Subscripts indicate the number of delays imposed on that sample. τ indicates 1 unit delay.

Taylor suggests three methods for detecting the presence of a fluctuating signal in the de-dispersed outputs. The first is a simple peak detector set at 3σ above the mean noise level. This would be adequate for sporadically strong pulsars like PSR 0531 + 21. For weaker, more steady pulsars a period search using the FFT or FFA method would be more appropriate. Since existing small computers are not capable of a real-time search for 32 channels simultaneously, the de-dispersed data must be stored on magnetic tape for off-line processing.

The third method for the detection process relies on a computation of the running variance of the de-dispersed outputs. A possible detection would be recorded when the running variance exceeds the off-pulse variance by some preset level. Taylor gives the integration time for a 3σ detection by this method as

$$T = 9\Delta t D^{-2} S^{-4},$$

where S is the peak strength of the signal being sought in units of the rms fluctuation of the de-dispersed outputs. For $\Delta t = 10$ msec, $D = 0.03$ and $S = 1$, then $T = 100$ sec. For signals with a higher duty factor such as the pulsar JP 1933 and other distant, as yet undetected pulsars whose pulses are severely broadened by interstellar scattering, this technique may be more satisfactory.

III. Dispersion

All cosmic radio signals are dispersed and scattered to some degree by propagation through the tenuous plasma of the interstellar medium, but only for time-varying signals can the dispersion be measured. Pulsar signals are the ideal probes to measure interstellar dispersion, allowing very accurate determination of the average electron density in the propagation path. However, dispersion distorts pulsar signals, limiting the time resolution of observations using conventional techniques. In this section, dispersion distortion and methods for the measurement and removal of distortion are discussed.

A. Transfer Function of the Interstellar Medium

We shall call the emitted pulsar signal $s(t)$. This is the voltage which would be induced in an antenna in the vicinity of the pulsar. For convenience in calculation we shall assume that the signal has a narrow bandwidth, which may be imposed at any point between the pulsar and the receiver detector. Then the pulsar signal can be written in the narrowband form:

$$s(t) = \text{Re}\,[v(t)\exp\,(i2\pi f_0 t)] \rightleftarrows S(f)$$
$$S(f) = \tfrac{1}{2}[V(f-f_0) + V^*(-f-f_0)], \quad (2)$$

where $\rightleftarrows$ indicates a Fourier transform. Thus,

$$v(t) = \int_{-\infty}^{\infty} V(f)\exp\,(i2\pi f t)\,df$$

and, where possible in this article, functions of time will be lower-case letters and functions of frequency will be upper-case. $v(t)$ is a slowly varying complex function, the complex envelope of the signal $s(t)$. $V(f)$ is centered at $f = 0$ and is nonzero only for $|f|$ less than a certain bandwidth. The signal $s(t)$ propagates through the interstellar plasma, which acts on $s(t)$ as a linear filter with a transfer function $H(f) = \exp[-i\kappa(f)z]$. z is the distance and $\kappa(f)$ is the propagation constant obtained from the dispersion equation for a tenuous plasma:

$$\kappa(f) = [1 - (f_p^2/f^2)]^{1/2} 2\pi f/c, \quad (3)$$

where the plasma frequency is given by $f_p^2 = N_e e^2/(4\pi^2 m_e \epsilon_0)$. N_e, m_e, and e are the electron number density, mass, and charge. In the interstellar medium $N_e \approx 0.03$ electrons per cubic centimeter, so $f_p \approx 2$ kHz, which is much less than the frequencies at which pulsars are observed and, therefore, Eq. (3) may be approximated by

$$\kappa(f) \approx [1 - (f_p^2/2f^2)] 2\pi f/c. \quad (4)$$

The interstellar medium transfer function is examined by expanding $\kappa(f)z$ in a Taylor series around the receiver center frequency f_0, giving the positive frequency part of $H(f)$ as

$$H_+(f) = \exp\left\{-i\left[\frac{2\pi z}{c}f_0\left(1 - \frac{f_p^2}{2f_0^2}\right) + \frac{2\pi z}{c}\left(1 + \frac{f_p^2}{2f_0^2}\right)(f-f_0)\right.\right.$$
$$\left.\left. - \frac{\pi z f_p^2}{c f_0^3}(f-f_0)^2 + \frac{\pi z f_p^2}{c f_0^4}(f-f_0)^3 + \cdots\right]\right\} \quad (5)$$

Since we are only interested in the narrow band of frequencies passed by the receiver, we can, without loss of generality, shift $H_+(f)$ to zero center frequency. The result, $H_+(f + f_0)$, is the Fourier transform of the complex impulse response $h_+(t)$ (Helstrom, 1960). Then

$$H_+(f + f_0) = \exp\,[-i(b_0 + b_1 f + b_2 f^2 + b_3 f^3 + \cdots)], \quad (6)$$

where $|f| \lesssim B/2 \ll f_0$.

The terms represent, respectively, a phase shift, a constant time delay, a time delay which depends linearly on frequency, and the curvature of the time delay vs. frequency.

The first term, $\exp(-ib_0)$, is an arbitrary phase shift which depends on z and f_p, and cannot be determined to within 2π. Furthermore, its effect upon the pulsar signal is lost upon square-law detection, so this term can be ignored.

Using the shift theorem (Bracewell, 1965),

$$X(f) \exp(-i2\pi f_0 t') \rightleftarrows x(t - t'), \qquad (7)$$

where $X(f)$ and $x(t)$ are a Fourier transform pair, we see that the second term represents the propagation time delay t_p for a signal envelope to travel to a distance z

$$t_p \equiv z/v_g = t_0[1 + f_p^2/2f_0^2], \qquad (8)$$

where the group velocity $v_g = 2\pi \, df/d\kappa$. Note that a signal of infinite frequency would arrive at the "light time," $t_0 = z/c$; lower frequency signals are delayed by $t_0 f_p^2/2f_0^2$, so that a pulse appears to sweep down in frequency at a rate

$$\alpha \equiv \frac{df}{dt} = -\frac{f^3}{t_0 f_p^2} = -\frac{f^3 N_e z e^2}{4\pi^2 c m_e \epsilon_0}. \qquad (9)$$

The quantity $N_e z$ is usually written as the integral of the electron density along the propagation path

$$DM = \int_0^z N_e \, dz, \qquad (10)$$

called the dispersion measure. Equation (8) can also be written as

$$t_p = t_0 + (D/f_0^2), \qquad (11)$$

where the term $t_0 f_p^2/2 \equiv D$ is the dispersion coefficient. Many observers now use D to specify dispersion in preference to DM, since D can be directly measured. Conversion to DM requires use of physical constants, some of whose values are not known exactly. Manchester and Taylor (1972) give the conversion factor as

$$DM(\text{pc cm}^{-3}) = 2.410000 \times 10^{-16} D(\text{sec Hz}^2). \qquad (12)$$

The second term in Eq. (6) can be eliminated by shifting the time origin of the impulse response $h_+(t)$ by t_p to the time when the pulse was received rather than the time it was emitted.

The third term (obtained from the derivative of the second) represents the linear rate of change of arrival time with frequency, the reciprocal of the frequency sweep rate α. Then

$$b_2 = \pi/\alpha = -2\pi D/f_0^3. \qquad (13)$$

The fourth term represents the second derivative, or the curvature, of the arrival time-vs.-frequency function, and

$$b_3 = 2\pi D/f_0^4.$$

Then Eq. (5), shifted to baseband, and with the time origin shift described above, converges to

$$H_+(f + f_0) = \exp\left[\frac{i2\pi D f^2}{f_0^2(f_0 + f)}\right], \quad f \ll f_0. \tag{14}$$

For many applications this may be approximated by the linear frequency sweep term only:

$$H_+(f + f_0) \approx \exp(i2\pi f^2 D/f_0^3) = \exp(-i\pi f^2/\alpha) \tag{15}$$

which, if f were unbounded, would have a Fourier transform:

$$h_+(t) \approx \exp(i\pi \alpha t^2)|\alpha|^{1/2} \exp(i\pi/4). \tag{16}$$

The accuracy of Eq. (15) relies on ignoring the f^3 and higher terms in Eq. (6). This is most critical near the band edges $f = B/2$, so that a criterion for Eq. (15) is $b_3(B/2)^3 < 1$ radian, which becomes $B < (8\alpha f_0/\pi)^{1/3}$. This condition is not always met for the bandwidths used in dispersion removal, in which case further terms or Eq. (14) must be used. A full expression, avoiding the assumption that $f_p^2 \ll f_0^2$ made in Eq. (4), shows that associated errors in b_2 and b_3 are typically 10^{-10} and can be ignored in all practical cases.

B. The Effect of Dispersion on Time Resolution

The transfer function or bandpass function of a receiver can be written in narrowband form as

$$E(f) = \tfrac{1}{2}[R(f - f_0) + R^*(-f - f_0)], \tag{17}$$

where $R(f)$ is the Fourier transform of the complex impulse response $r(t)$ of the receiver. The phase and amplitude response of the receiver system, given by $R(f)$, is centered on $f = 0$ and is zero for frequencies outside $\pm B/2$, where B is the receiver bandwidth.

After propagation through the interstellar medium the Fourier transform of the received signal is

$$S(f) \exp[-i\kappa(f)z]\, E(f) = \tfrac{1}{8}[V(f - f_0) + V^*(-f - f_0)][H_+(f) + H_-(f)][R(f - f_0) + R^*(-f - f_0)], \tag{18}$$

where $V(f)$ is given by Eq. (2) and $H_+(f)$ is given by Eq. (14). The two parts of V, H, and R correspond to positive and negative frequencies. The resultant complex envelope $y(t)$ has a Fourier transform which is just the product of these functions shifted by f_0, and thus $y(t)$ is a double convolution (denoted by $*$) in the time domain.

$$y(t) = v(t)*h_+(t)*r(t) \rightleftarrows V(f)H_+(f+f_0)\,R(f). \tag{19}$$

That is, $y(t) = v(t)*r_\alpha(t)$, where $r_\alpha(t)$ is the net impulse response for voltage given by

$$r_\alpha(t) = \int_{-\infty}^{\infty} R(f)\exp(-i\pi f^2/\alpha + i2\pi ft)\,df. \tag{20}$$

If the bandpass is rectangular, $r_\alpha(t)$ can be expressed in terms of Fresnel integrals (e.g., Klauder et al., 1960; Cole, 1972). More illuminating, however, is to evaluate Eq. (20) by the stationary phase method:

$$r_\alpha(t) \approx \alpha^{1/2} R(\alpha t)\exp(i\pi\alpha t^2 - i\pi/4), \tag{21}$$

which is valid provided that $R(f)$ is a smoothly varying response of bandwidth $B \gg \alpha^{1/2}$.

The output of a square-law detector following the receiver is $I(t) = |y(t)|^2/2$ (providing that the postdetection filter removes frequencies near $\pm 2f_0$). We now consider the time resolution for practical intensity observations.

$$I(t) = \tfrac{1}{2}|v(t)*r_\alpha(t)|^2. \tag{22}$$

Of most importance is the special case where the emitted pulse is incoherent [i.e., successive samples of $v(t)$ are independent, as for amplitude-modulated white noise]. Applying this idea under some postdetection smoothing and using Eq. (21), we obtain

$$I(t) = \frac{1}{2}\int_{-\infty}^{\infty}|v(t_1)|^2\,|R(\alpha t - \alpha t_1)|^2 \alpha\, dt_1. \tag{23}$$

We see that the pulsar intensity envelope is convolved with a resolution function, which is the receiver power response converted to a time function by the pulsar sweep rate α. An emitted impulse would then be detected as a pulse of duration

$$t_s = B/|\alpha| \qquad \text{for} \quad B \gg |\alpha|^{1/2} \tag{24}$$

or, if $B/|\alpha| < 1/B$, the filter rise time $1/B$ will determine the pulse duration. In principle, if the passband were rectangular, the impulse response should show Fresnel pattern ripples as in a lunar occultation record. However, with practical bandshapes and postdetection smoothing, the ripples will be washed out.

Thus the time resolution for conventional pulsar observations is limited to t_s or $1/B$, whichever is greater. Resolution can be optimized by narrowing the receiver bandwidth until

$$t_s = t_{\min} = 1/B = |\alpha|^{-1/2}, \qquad (25)$$

but at the expense of reducing the signal-to-noise ratio. To achieve this resolution, the detector time constant must be reduced to $1/B$, but then, from the radiometer equation, we see that the random fluctuations of the detector output power, due to estimation error, become comparable with the mean detector output (see Section VI).

At low frequencies and for high-dispersion pulsars, the dispersion degradation of time resolution becomes serious. For example, the best time resolution that can be obtained for even the lowest dispersion pulsar PSR 0950+08 at 111.5 MHz is 134 μsec.

C. DISPERSION REMOVAL

1. *Postdetection Dispersion Removal*

In conventional radioastronomy observations, the signal-to-noise ratio can be improved by using wider receiver bandwidths. But we have just seen that for pulsar signals this leads to increased dispersion smearing, and hence to poorer time resolution. By dividing the receiver band into many narrow channels and then adding the detected outputs after appropriate delays, the time resolution of a single narrow bandwidth channel can be retained, with the signal-to-noise ratio equivalent to that attainable for the total rf bandwidth. Figure 2b demonstrates the principle of postdetection dispersion removal or "signal enhancement." In the simplest case the detected outputs are sampled at intervals of $t_s = 2BD/f_0^3$, where B is the bandwidth of an individual channel centered at frequency f_0. Let x_{ij} be the sampled output of channel i ($i = 1, N$) at time j. Then the channel outputs are summed according to

$$y_j = \sum_{i=1}^{N} x_{i,j-i},$$

providing a signal-to-noise ratio enhancement of $N^{1/2}$, if the adjacent filter bandwidths are independent (Boriakoff, 1973) and the off-pulse noise is Gaussian. This process is easily implemented in software (Craft, 1970), and several devices have been built to perform the operations in real time. Orsten (1970) constructed an analog delay line with discrete delay stages using the bucket-brigade principle (Sangster and Teer, 1969). Boriakoff (1973) has built a digital pulsar processor which can be oper-

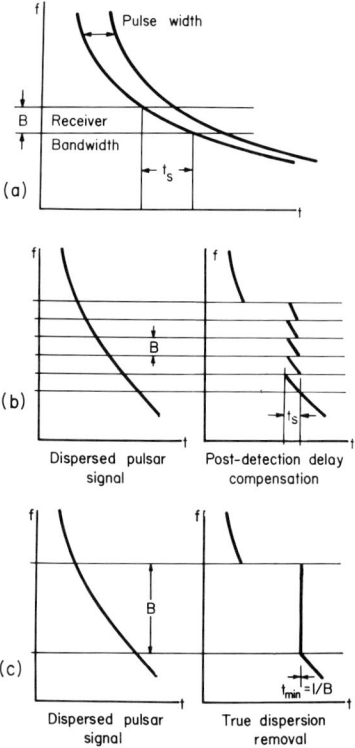

FIG. 2. (a) A diagrammatic representation of a dispersed pulsar signal whose frequency decreases with time. Temporal structure in the signal is smeared over the bandwidth sweep time t_s. (b) By dividing the receiver band into narrower frequency channels the time resolution can be improved (or the signal-to-noise ratio enhanced over that for a single narrowband channel) by appropriately delaying the signals from each channel and summing the detected outputs. (c) In a true dispersion removal process the delay-vs.-frequency characteristic varies smoothly across the receiver passband so that all the energy from a signal component, represented here by a dispersed "impulse," is compressed to a time interval of $1/B$.

ated either as a 32-channel pulsar signal enhancer or as a spectrum analyzer. Both of these devices have had extensive use, but they have the same fundamental minimum time resolution as a single-channel receiver, given in Eq. (25).

2. Predetection Dispersion Removal

For studying the fine detail in pulse structure the time resolution can be improved considerably by removing dispersion distortion before detection. From Eq. (18) we see that, in principle, the emitted signal can be recovered over a limited frequency and time interval simply by passing the received signal through the inverse filter whose transfer function is $\exp[+i\kappa(f)z]$. The time resolution is then limited to the width of $r(t)$ i.e., $1/B$, and the practical resolution limit is set by the widest bandwidth which can be sampled sufficiently fast to avoid aliasing, as compared with the narrowest practical bandwidth for the postdetection delay compensation technique. Predetection dispersion removal can be achieved either by using a hardware dispersive filter or by simulating the filter by computer software. Both methods have their distinct advantages and disadvantages, particularly in the tradeoff between the speed of a hardware device and the versatility of a software process.

A software predetection dispersion removal technique which has been used extensively for studies of pulsar microstructure (Hankins, 1971, 1972, 1973) is described below.

The received pulsar signal is split into two identical parts and mixed to baseband by multiplying by $\cos 2\pi f_0 t$ and by $\sin 2\pi f_0 t$. After filtering out the terms centered on $2f_0$, the two products are digitized and combined in a complex word that represents the complex envelope as received, given by Eq. (19). The complex samples are recorded on magnetic tape for off-line processing.

The software filtering process takes advantage of the fast Fourier transform techniques of digital filtering (Cooley and Tukey, 1965; Sande, 1968). We denote functions of time by lower-case letters x_j and their discrete Fourier transform by upper-case letters, for example,

$$X_k = F[x_j] \equiv \frac{1}{N} \sum_{j=0}^{N-1} x_j \exp(i2\pi jk/N).$$

Due to the periodic nature of the transform, certain precautions must be taken (Bergland, 1969). In the general case, we want to perform the discrete convolution

$$y_i = \frac{1}{N} \sum_{j=0}^{N-1} x_{i-j} h_j \tag{26}$$

in the transform domain such that

$$Y_k = X_k H_k,$$

where H_k and h_j represent sampled versions of Eq. (14) and its Fourier transform. If X_k and H_k are computed directly using the FFT, the resultant product, transformed back to the time domain, will be the cyclical convolution, since the FFT treats both X_k and H_k as if they were periodic. The problem is easily side-stepped, however, in the following way. The impulse response h_j containing $N < 2^{m-1}$ samples is extended with zeros by defining

$$\hat{h}_j = h_j, \qquad 0 \leq j < 2^{m-1};$$
$$\hat{h}_j = 0, \qquad 2^{m-1} \leq j < 2^m.$$

Then compute X_k and H_k via the FFT of length 2^m, and compute the product $Y_k = X_k H_k$, then find $y_i = F^{-1}[Y_k]$ via the FFT, where F^{-1} indicates reverse transform. The first 2^{m-1} lags of y_i are meaningless, and the second 2^{m-1} lags of y_i constitute the desired convolution for $0 \leq i < 2^{m-1}$. The next 2^{m-1} lags of y_i are obtained by moving 2^{m-1} samples further along the data series y_i, and repeating the process, as summarized in Fig. 3.

The filter transfer function from Eq. (14) is computed from measured values of D. Its shape is shown in Fig. 4.

The effectiveness of the dispersion removal filter is shown in Fig. 5 where an idealized dispersed impulse has been filtered. The central peak is only one sample interval $(1/B)$ wide and the "side lobes" are very low. Figure 5 also shows that, to recover dispersed impulses whose intrinsic width is $1/B$ or less, the dispersion coefficient D must be known to within about $2/Bt_s$. An error ΔD in the dispersion coefficient results in a residual time smearing $\Delta t_s = 2B\Delta D/f_0^3$ but, if the time constant of postdetection smoothing is significantly greater than Δt_s, then the error ΔD can be ignored.

There are two approaches to achieve real-time predetection dispersion removal: first, to speed up the software technique described above by using dedicated hardwired FFT processors; and second, to perform the filtering by convolution in the time domain using some form of transversal filter (Squire et al., 1969). A transversal filter consists of a tapped delay line along which the signal is propagated, coefficient multipliers at each tap, and an adder to combine the products. One possible configuration is shown in Fig. 6, a realization of the discrete convolution given by Eq. (26). Such a filter could be built using entirely digital circuitry (Hankins, 1974b; M. A. Feyjoo, private correspondence, 1974), by utilizing analog charge-coupled devices (IEEE Intercon, 1974), by tapped surface wave delay lines, or by an optoacoustical processor (Squire et al., 1969).

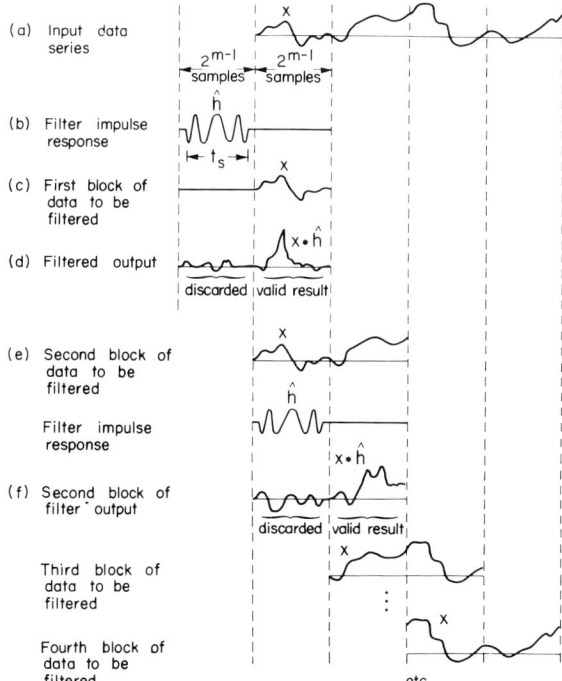

FIG. 3. (a) The data series x is divided into blocks of 2^{m-1} samples. (b) The band-limited impulse response of the dispersion removal filter h has a duration t_s. It is extended with zeroes to form $\hat{h}$ of length 2^m samples to avoid the periodic replication imposed by the FFT. (c) The first 2^{m-1} samples of x are preceded by 2^{m-1} zeroes. Then the 2^m-point FFT of this block is multiplied by the FFT of $\hat{h}$. (d) Only the second half of the reverse transform of the product is valid and retained. The first half is discarded. (e) The block pointer is advanced 2^{m-1} samples and the process is repeated to give the next 2^{m-1} output values, shown in (f).

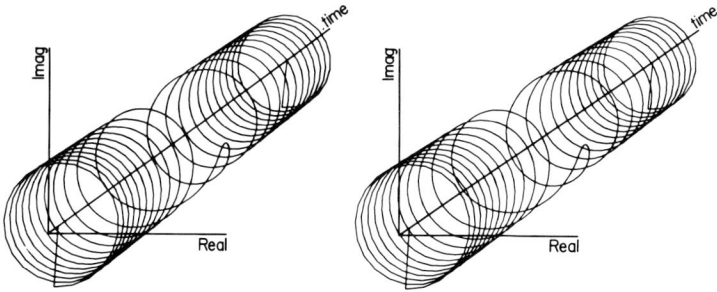

FIG. 4. Impulse response $h(t)$ of the dispersion removal filter, plotted in oblique projection as a stereo pair. The duration of the function is the sweep time t_s across the receiver bandwidth.

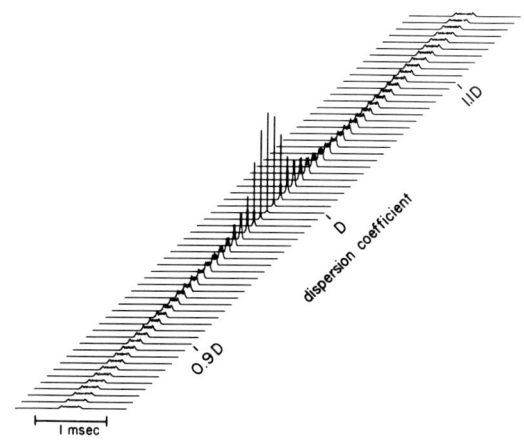

AMBIGUITY DIAGRAM

Fig. 5. Dispersion removal ambiguity function, plotted with time as the independent variable. Each trace is the convolution of an idealized noise-free pulse dispersed by a dispersion coefficient D, with the impulse response of a dispersion removal filter whose dispersion constant is the fraction of D noted on the right of the figure. When the dispersions are perfectly matched, the impulse is recovered with a peak-to-sidelobe ratio of 67. The convolutions have been successively shifted to the right for clarity. The time resolution is equivalent to 8 μsec over a simulated bandwidth of 125 kHz. The total energy of the filter output is the same whether or not the dispersion of the filter is matched to the dispersion of the filter, but it is rearranged in time.

Dispersive lumped-constant longitudinal filters (O'Meara, 1960) could also be used, but separate filters would be required for each dispersion measure, each frequency, and each bandwidth desired—clearly an impractical project for more than a very few DM, f_0, B combinations.

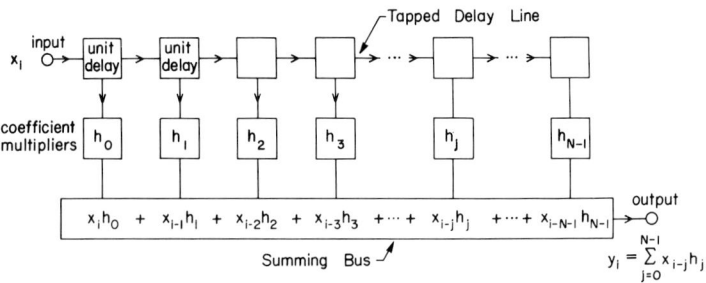

Fig. 6. Transversal filter. The input signal is sampled at equal delay intervals along the delay line. The samples are weighted by the sampled values of the filter impulse response h_j, and then summed to form the discrete convolution.

3. Comparison of Predetection and Postdetection Dispersion Removal Techniques

The principal advantage of the predetection dispersion removal technique is the high time resolution which can be attained at low frequencies and large dispersions. The time resolution is limited only by the widest bandwidth that can be sampled. Implementation of the filter in software using a typical computer with 3-μsec operation time requires as much as 1000 times longer to process the data than it does to record it. This is no handicap for high time resolution studies of a few thousands of pulses, since secondary analyses and processing require more computer time than the dispersion removal; but the software technique is impractical for routine synoptic pulsar studies. Using a software filter, however, imperfect receiver characteristics, instrumental and ionospheric polarization can be compensated for during the filtering. Furthermore, replication of filters for multiple frequency and polarization measurements is trivial in software, but expensive in hardware.

The postdetection dispersion removal technique, even in its simplest configuration, where the delays and additions are done in software, requires a more complicated receiver (multichannel filter bank). However, since the computations are simple, they can be done in real time. At frequencies above 800 MHz, good time resolution can be obtained for most pulsars even when the individual filters in a filter bank are quite wide. To optimize time resolution, however, filters of different widths are required for each dispersion measure–frequency combination desired. These filter banks tend to be expensive and tedious to maintain.

4. Swept Local Oscillator Methods

The usefulness of both predetection and postdetection dispersion removal can be enhanced by sweeping the receiver local oscillator at intervals of P_1 at a rate $df_0/dt \leq \alpha$, where α is given by Eq. (9). This causes the pulsar to remain in the receiver band for a longer time, effectively increasing t_s and B. If the local oscillator sweep rate matches α, then, calling the IF output the real part of $c(t)$, it can be shown that its spectrum $C(f)$ is given by

$$C(f) = v(f/\alpha) \exp(-i\pi f^2/\alpha + i\pi/4)|\alpha|^{1/2}. \tag{27}$$

This result involves a stationary phase integral which requires that the total bandwidth $B \gg |\alpha|^{1/2}$ (i.e., there must be substantial dispersion smearing across the bandwidth, which is the case of interest). The emitted envelope limited by the receiver band, $v(t)$, is then translated into the frequency domain according to the sweep rate α. The power spectrum

$|C(f)|^2$ would thus yield $|v(f/\alpha)|^2$, the intensity *without* dispersion. Splitting the IF into a multichannel spectrometer with frequency resolution Δf thus yields the de-dispersed signal with time resolution $\Delta f/|\alpha|$ (e.g., Sutton et al., 1970). This is the same as that obtainable using a predetection bandwidth Δf. The important difference is that the output of each channel is averaged over the total sweep time $B/|\alpha|$. Consequently, we are only constrained that $\Delta f \geq |\alpha|/B$, allowing equivalent time resolution down to near $1/B$, in contrast to the limit of Eq. (25). For predetection processing, it appears that a valuable increase in effective total bandwidth can be achieved without increasing the data rate to equal that bandwidth. The practical snag is that a very large number of channels ($B/\Delta f$) would be required—for example, about 1000 for a 0.1 msec resolution across 1 MHz for a medium-dispersion pulsar.

A variation of this process has been used by several observers (Staelin and Sutton, 1970; Hankins, 1974a) for observing "giant" pulses from the Crab pulsar PSR 0531+21. The intrinsic main pulse is so short (200–400 μsec) that the pulse can be received at one frequency, then the local oscillator switched to a lower frequency, and the same pulse can be received again. In this way Hankins sampled "giant" pulses in 20 bandwidths of 125 kHz each, spanning 437 to 422 MHz. When added together after predetection dispersion removal from each channel, an effective bandwidth of 2.5 MHz was obtained. Figure 7 shows one giant pulse obtained this way.

The somewhat analogous technique for removing dispersion from interferometer observations of pulsars is described by Erickson et al. (1972).

5. Observations Using Dispersion Removal Techniques

As noted by Taylor (1973), the ideal pulsar receiver would always include a dispersion removal system. The increasing use of such techniques has led to many new results in recent years. Taylor and Huguenin (1971) applied postdetection dispersion removal to total bandwidths of up to 40 MHz in a study of individual pulses from a large sample of pulsars. They were able to include weak pulsars which are only detectable as averages using conventional techniques and made extensive observations of their pulse-to-pulse fluctuations. Backer et al. (1973), using postdetection processing, discovered weak interpulses and emission spanning a large fraction of the period in the average profile of several pulsars. Evidently, improved time resolution and signal-to-noise ratios are achievable in timing and intensity variation measurements, if dispersion removal can be used.

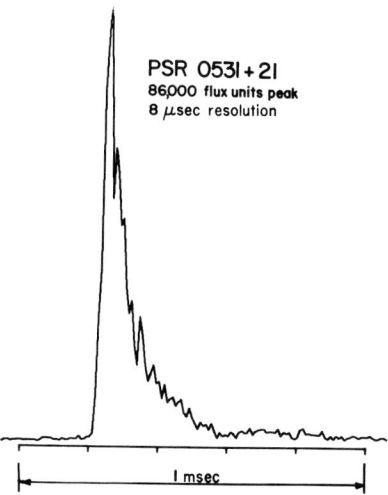

Fig. 7. A single "giant" pulse from the Crab pulsar. Dispersion was removed from 20 dual-polarization 125-kHz channels spaced from 422 to 437 MHz. Then the detected outputs were added together to yield a 2.5-MHz effective bandwidth. Zero detector level is shown by the base line. The Crab Nebula background level is shown to the left of the pulse. The effect of receiver noise is negligible. The rise time is less than 50 μsec. The approximately exponential "tail" may be due to interstellar scattering.

The predetection dispersion removal technique developed by Hankins (1971) immediately revealed short, strong, isolated micropulses of duration less than 10 μsec and peak intensities of more than 35,000 flux units (1 F.U. = 10^{-26} W/m²/Hz) in the signals at 111 MHz from PSR 0950+08 and PSR 1133+16. These signals imply an equivalent source brightness temperature of over 10^{30}°K and extremely small emitting regions, placing severe limits on the emission mechanism (Manchester et al., 1973). Very narrow features ("notches") were also revealed in the average profile of PSR 1919+21 below about 200 MHz, which were only detectable after removing dispersion (Hankins, 1972).

D. Measurement of Dispersion

1. Methods

In practice, dispersion is measured by recording the times of arrival (t_1, t_2) of a pulsar signal at two or more widely separated frequencies (f_1, f_2). Then, from Eq. (11) the dispersion coefficient D is obtained.

$$t_1 - t_2 = D(1/f_2^2 - 1/f_1^2). \tag{28}$$

There have been several refinements and corrections to the basic technique; these have been discussed in the literature and are summarized below.

The delay in time of arrival between two different frequencies can be determined for a single pulse if the pulse is strong enough to be seen clearly on both frequencies. To maximize the delay, the highest and lowest practical frequencies should be used. But most pulsars are weak at low frequencies, so time-of-arrival measurements are usually performed on the averages of many pulses to reduce uncertainties due to noise. Then it is the relative phase of the pulse trains at the various receiver frequencies that is actually measured. The techniques and accuracy of such measurements are discussed in Section IX. Since it may take several pulse periods for a pulse to drift down from one receiver frequency to another, say an octave lower, the actual time delay is this phase difference plus an *a priori* unknown number of pulse periods. The pulse period ambiguity must be resolved by using narrower frequency separations or other information.

Average pulse profiles have advantages over single-pulse measurements, giving improved signal-to-noise ratio and reduction of frequency-selective fading from interstellar scintillation. To optimize the timing accuracy for dispersion measurements, the frequency separation should be as wide as possible, maximizing the delay. However, nearly all pulsars have frequency-dependent pulse shapes, and it is difficult to decide which features of the average pulse profile correspond over wide ranges of frequency. Craft (1970) found that, for pulsars with complex average profiles, the time delay between the centroids of the average profiles at each frequency must be used to give results consistent with the dispersion law [Eq. (4)]. Alternatively, it may be argued that the sharpest feature in the profiles should be used—for example, the narrow notches already mentioned for PSR 1919+21 provide a useful fiducial mark on the profile. If the notch longitude were a function of radio frequency this would, of course, create an unavoidable error in the resultant value of D.

Yet another technique is possible, which involves the average cross-correlation function of individual pulses between widely spaced frequencies. This is a refinement of the technique using single pulses, and is possible because the subpulse details are often correlated over a wide frequency range. Indeed, Rickett (1974) found that even the microstructure in PSR 0950+08 was highly correlated between 318 and 111.5 MHz. Figure 8 shows the cross-correlation function between the 111 and 318 MHz intensity records, each of which was filtered to remove dispersion across the individual receiver bands. The time shift of the peak in the correlation can be measured to an accuracy of 50 μsec, which could be improved to 10 μsec by model-fitting the cross-correlation function.

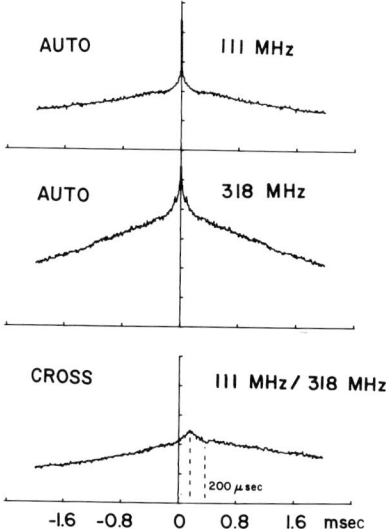

FIG. 8. Intensity, autocorrelations, and cross correlation of 111.5- and 318-MHz records from PSR 0950+08. The short time scale feature in the autocorrelation functions also appears in the cross correlation indicating that microstructure is wideband, permitting precise determination of the dispersion coefficient.

This gives the possibility of dispersion measurements accurate to 1 part in 10^5. Until now, such accuracies have only been possible for the Crab pulsar.

2. Corrections to Measurements of the Dispersion Coefficient

(a) Instrumental corrections for the receiver bandpass and detector characteristics are discussed by Rickett (1969), Craft (1970), and Boriakoff (1973); these are summarized in Section IV. The response of the IF amplifier filter to a dispersed signal, the delays imposed by the postdetector time constant and the data recording system must all be compensated for.

(b) The motion of the observer causes an apparent Doppler shift of the delay interval $t_2 - t_1$ and the received frequencies, f_2 and f_1, which requires the correction $D = D'(1 - v/c)$, where D' is the observed dispersion coefficient, D is corrected to the solar system barycenter, and the radial velocity v of the observer is positive toward the plasma through which the observer is moving. Since the orbital velocity of the Earth $v/c \leq 10^{-4}$, this correction is necessary for measurements made with greater accuracy than 1 part in 10^4, such as those made routinely by Rankin and Counselman (1973) for the Crab pulsar PSR 0531+21.

(c) The integrated electron content of the ionosphere is less than 10^{-6} times that of the interstellar medium and therefore can usually be ignored. However, the interplanetary electron contribution to the dispersion delay can amount to several milliseconds at 100 MHz for propagation paths passing through the corona closer than 10 solar radii from the sun. In fact, Counselman and Rankin (1972) have measured the density of the solar corona using the dispersion delay of pulses from the Crab pulsar, which is occulted by the sun annually on or about June 15.

(d) Interstellar scattering broadens the pulses from some pulsars so much that it dominates the pulse shape and width. The scattering is strongly frequency-dependent: thus, a delay and pulse width proportional to f^{-4} are added to the f^{-2} delay of interstellar dispersion. Counselman and Rankin (1971) developed a technique for extracting both the scattering and dispersion coefficients from the Crab pulsar. They measured the phases of the Fourier coefficients of the average profile (see Section VIII), and by a least-squares-error fitting procedure extracted both parameters.

(e) The accuracy with which f_1 and f_2 in Eq. (28) are known is also important for accurate dispersion measurements. The centroids of the receiver bandwidths must be known to as good a percentage accuracy as the time difference. If there are narrow scintillation features across the pulse spectrum (Section VIII) they could shift the effective center frequency by as much as half the receiver bandwidth, causing very significant errors.

3. Accuracy of the Tenuous Plasma Dispersion Law

The work of Tanenbaum et al. (1968), Drake and Craft (1968b), Craft (1970), and others has shown that the dispersion law given by Eq.(4) correctly represents the measurements to within about 1 part in 10^4; plots of $t_2 - t_1$ versus $(1/f_2^2 - 1/f_1^2)$ for several combinations of frequencies can be fit by a straight line. If Eq. (4) were an inadequate representation of the dispersion law, the plots would curve toward earlier or later arrival times. Modifyng Eq. (4) to include the first-order effect of a weak longitudinal magnetic field and the second-order approximation from $f_p^2 \ll f^2$, then from Craft (1970), Eq. (28) becomes

$$t_2 - t_1 = B(1/f_2^2 - 1/f_1^2)(1 + T_1 + T_2),$$

where

$$T_1 \approx 60.8 \int_0^Z N_e^2 \, dz / f_2^2 \int_0^Z N_e \, dz$$
$$T_2 \approx 5.6 \times 10^{10} \int_0^Z N_e B \cos \theta \, dz / f_2 \int_0^Z N_e \, dz.$$

The expressions are given in MKS units and apply for $f_2 < 0.3\ f_1$. $B \cos \theta$ is the component of the magnetic field in the propagation direction. The degree to which the measurements indicate T_1 and T_2 can be neglected can be used to obtain limits on the plasma density and longitudinal magnetic field. Craft (1970) obtained an upper limit for N_e of about 10^4 cm^{-3}, and a maximum longitudinal magnetic field of about 5×10^{-3} G for five pulsars. The electron density upper limit provides a lower limit for the extent of the dispersing region of about 3×10^{-4} pc. Since all the pulsars measured give approximately the same limits though their dispersions are quite different, it is highly unlikely that the dispersion is so localized. The magnetic field upper limit is some orders of magnitude higher than that actually measured by the Faraday rotation measurements of Manchester (1972) and others.

IV. Sampling, Resolution, and Average Profiles

A. Instrumental Considerations

The basic observable quantity in radio studies of pulsars is the varying voltage induced in a certain receiver–antenna combination. The center frequency f_0 and bandwidth B of the receiver, together with the polarization response of the antenna, determine the instrumental effects. We denote the voltage $s(t)$ and define the associated complex envelope function $v_1(t)$ by

$$s(t) = \mathrm{Re}[v_1(t) \exp(2\pi i f_0 t)]. \tag{29}$$

In an ideal experiment we would also record the output of a similar receiver–antenna combination with (hopefully) orthogonal polarization. From the two complex signals v_1 and v_2, we could construct four independent real parameters to describe the varying polarization (e.g., the Stokes parameters). Polarization techniques are discussed in Section V. We consider now the detection and recording of the pulses from a single antenna–receiver combination.

In conventional observations the voltage $s(t)$ is amplified, detected and smoothed, giving $I(t) = s^2(t)*k(t)$, where $k(t)$ is the impulse response of the postdetection filter, and "$*$" means a convolution. The filter removes the components of $s^2(t)$ at $\pm 2f_0$ and also determines the time resolution. It follows that $I(t) = \frac{1}{2}|v(t)|^2 * k(t)$. The intensity signal is then sampled at an interval Δt_s which is determined by the postdetection filter. In order to avoid aliasing, the Nyquist frequency, $1/2\Delta t_s$, must be above the filter cutoff frequency. Filters cutting off abruptly in the frequency domain are economical from the point of view of sampling but tend to have an "overshoot" in their impulse response, producing artificial ripples in the data. Most pulsar observations have employed a simple RC circuit which requires more frequent sampling compared to its time resolution, but which has a monotonic impulse response (truncated exponential) giving smoother looking pulses.

An alternative sampling method has sometimes been used for very fine time resolution. The signal $s(t)$ is mixed to baseband (multiplying by $\cos 2\pi f_0 t$ and $\sin 2\pi f_0 t$). After the outputs are filtered to eliminate the components at $\pm 2f_0$, the *voltages* are sampled. Since the two channels are cut off outside $\pm B/2$ each channel must be sampled at $\Delta t_s \leq 1/B$, giving at least $2B$ samples per second. Only a few observatories have sampling fast enough to observe in this fashion (for example, at Arecibo, data rates up to 250 kHz can be used). The sine and cosine samples are the real and imaginary part of $v(t)$, the complex envelope function. In this method the detection and smoothing are done by software. The advantages of this predetection sampling include the possibility of very fine time resolution by coherent dispersion removal (Section III), and the flexibility of varying the time and frequency resolutions in subsequent off-line analysis (Section VII).

Either of the above methods leads to data recorded as a series of sampled intensities I with a certain time resolution. For normal (postdetection) sampling the signal $|v(t)|^2$ is convolved with a net resolution function, $w(t)$, as follows:

$$I(t) = \tfrac{1}{2}|v(t)|^2 * w(t)$$
$$w(t) = g(t) * |R(\alpha t)|^2 * k(t), \qquad (30)$$

where $g(t)$ is the impulse response of the interstellar scattering (Section VIII); $|R(\alpha t)|^2$ is the receiver bandpass converted into a time function by the dispersion sweep rate [Section II, Eqs. (9) and (23)]. The net time resolution τ_r is the width of $w(t)$, dominated by the longest of the three contributing responses. Since system noise is only subject to the final smoothing, the best signal-to-noise ratio is obtained by adjusting the postdetection time constant τ to match the dispersion sweep time. For predetection recording the dispersion can be removed coherently, followed by software detection and smoothing, allowing time resolution down to $1/B$ (or the interstellar scattering time constant, whichever is longer). We denote the net resolution by τ_r and the rms system noise level by σ_N.

B. Average Profiles

The sampling time base may or may not be synchronized with P_1, the pulsar period. For synchronous sampling, the average profile is computed directly by summing the intensities in each of the $P_1/\Delta t_s$ resolution cells of the period. Evidently, $P_1/\Delta t_s$ must be an integer or the sampling must be synchronized afresh each period. Several standard hardware devices exist for this signal averaging, going under various names—pulse height analyzer, multiscaler, etc. The resultant average profile is smeared by an additional resolution function if the assumed period P_0 is different from the true period P_1 (at the observatory). For N periods averaged, the additional smearing function is a rectangle of width equal to $N|P_1 - P_0|$. Over the course of a day the rotation of the Earth changes the apparent period. This can be viewed as a $21.2 \sin E$ millisecond difference in arrival time between the observatory and the center of the Earth, changing as the elevation angle E changes. Alternatively, it can be regarded as a changing Doppler shift equal to $0.0212 \cos E \; dE/dt$ as a fraction of the geocentric period. In either notation, the net time smearing in an average profile is the accumulated timing difference between the pulses as measured at the observatory rotating with the Earth and the assumed period. Most observatories have developed pulsar period ephemerides which account for Earth spin and orbital rotation and also secular increases in the intrinsic period (see Section IX). Time standards synchronized with UTC (or some other atomic time) to 10 μsec have been developed. In some observations the individual samples from each period are stored (e.g., on magnetic tape), in which case adjustments to the assumed period can be inserted later, but clearly this leads to enormous quantities of data and more processing.

The second sampling scheme is asynchronous, when $P_1/\Delta t_s$ is not an integer. The algorithm for computing the mean profile thus involves a decision into which phase interval of the assumed period should each sample be added. If $t_n = n\Delta t_s$ is the time of the nth sample, the appropriate phase is the fractional part of t_n/P_0, quantized to the nearest resolution cell. If the period is divided into cells of width Δt_p, the resultant profile is of course smeared over Δt_p in addition to the smearing over the resolution time τ_r. There is thus some resolution advantage in dividing the period into cells smaller than Δt_s; indeed, early measurements at Jodrell Bank took the process one step further by deliberately undersampling the data (e.g., $\Delta t_s = 10$ msec with $\tau_r = 2$ msec), but matching Δt_p, the duration of the profile cell, to the time resolution τ_r. Since the sample interval and period were not commensurate, over the course of $\Delta t_s/\Delta t_p$ periods a sample would be sprinkled into each cell in the profiles.

This technique relies on the profile remaining stationary and requires $\Delta t_s/\Delta t_p$ times as many pulse periods to achieve the same confidence level in the average profile. Nevertheless, in the early stages of pulsar investigations it was used successfully since rapid data sampling was not then available (e.g., Davies et al., 1968).

V. Polarization

Astronomers generally agree that pulsars are rotating neutron stars with strong magnetic fields. The observation that pulsar radio signals are often highly polarized supports this hypothesis. In fact, the polarization characteristics of pulsars are among the most critical observations in the development of satisfactory pulsar radiation models.

The techniques used for pulsar polarization measurements are not unique in radio astronomy. However, the complex signal approach necessary for the predetection dispersion removal technique, discussed in Section III, can be used to advantage in pulsar polarization measurements also. We show here only the essential differences that arise between standard methods and the complex signal approach. A complete treatise on polarization measurements would require a separate paper by itself.

For polarization measurements two concentric antennas with approximately orthogonal linear or circular polarizations are used. The voltages induced in these antennas are real and can be represented as $X(t)$, $Y(t)$ or $R(t)$, $L(t)$ for nominally linear or circular pairs. $X(t)$, for example, can be written as a narrowband signal

$$X(t) = \text{Re}\{x(t) \exp(i2\pi f_0 t)\}.$$

The Stokes parameters are then obtained as

$$I = \langle X^2 \rangle + \langle Y^2 \rangle = \langle R^2 \rangle + \langle L^2 \rangle$$
$$Q = \langle X^2 \rangle - \langle Y^2 \rangle = 2\langle RL \rangle$$
$$U = 2\langle XY \rangle = 2\langle RL' \rangle$$
$$V = 2\langle XY' \rangle = \langle R^2 \rangle - \langle L^2 \rangle,$$

where $Y' = Y(t + \frac{1}{4}f_0)$ and $L' = L(t + \frac{1}{4}f_0)$ are signals delayed by a quarter wavelength. In conventional radio astronomy observations, analog radio frequency circuits are used to generate the various linear combinations of X, Y, and Y', and the multipliers are achieved by phase switching techniques and crystal detectors. The outputs are passed through postdetection low-pass filters to perform the time averaging indicated by the angular brackets. In pulsar observations phase switching is not used, since convenient switching times are similar to the desired time resolutions. This makes the calibration of the polarimeter more critical, since without phase switching, the quantities I, Q, U, and V include dc contributions.

Two groups, Campbell *et al.* (1970) and Hankins (1974a), have taken advantage of the rapid sampling capabilities at the Arecibo Observatory to record the complex envelopes $x(t)$ and $y(t)$ or $r(t)$ and $l(t)$ directly. At this point, if desired, the time resolution can be improved by applying the predetection dispersion removal technique of Section III to both polarization channels. Differential Faraday rotation across the receiver band can also be removed at this point, by processing the right- and left-hand circular signals with different dispersion coefficients. Taking only the quadratic term in Eq. (6), corrections to the coefficient b_2 [Eq. (13)] should be

$$\Delta b_2 = \pm 0.18 f_0^{-4} \int NB \cos\theta \, dz \quad \text{(MKS units)}$$

for the right and left circular polarization channels.

The sampled representations of the complex envelope are combined in the computer to form the Stokes parameters.

$$\begin{aligned} I &= \langle |x|^2 \rangle + \langle |y|^2 \rangle = \langle |l|^2 \rangle + \langle |r|^2 \rangle \\ Q &= \langle |x|^2 \rangle - \langle |y|^2 \rangle = \langle \mathrm{Re}(r^*l) \rangle \\ U &= \langle \mathrm{Re}(x^*y) \rangle = -\langle \mathrm{Im}(r^*l) \rangle \\ V &= \langle \mathrm{Im}(x^*y) \rangle = \langle |l|^2 \rangle - \langle |r|^2 \rangle \end{aligned}$$

A. Polarimeter Calibration

The vector complex signal $\mathbf{s}(t)$ appearing at the terminals of the nominally orthogonally polarized antennas are related to the complex signal voltage $\mathbf{s}'(t)$, which would be induced in a perfect polarimeter, by a 2×2 complex matrix $\mathbf{A}$ which describes the true polarization response of the antenna.

$$\mathbf{s}(t) = \mathbf{A}\mathbf{s}'(t)$$

where, for example, $\mathbf{s}(t)$ may be

$$\mathbf{s}(t) = \begin{pmatrix} x(t) \\ y(t) \end{pmatrix} \quad \text{or} \quad \begin{pmatrix} r(t) \\ l(t) \end{pmatrix}.$$

Then the true Stokes parameters of the field in the antenna vicinity can be obtained from

$$\mathbf{s}'(t) = \mathbf{A}^{-1}\mathbf{s}(t).$$

The antenna calibration matrix **A** can be determined by transmitting a signal at the frequency f_0 from a concentric test dipole rotating at Ω radians per second. If sufficient care is used in constructing the test dipole, then the circularly polarized components of its transmitted field are

$$\mathbf{s}'(t) = \begin{pmatrix} \cos \Omega t - i \sin \Omega t \\ \cos \Omega t + i \sin \Omega t \end{pmatrix}.$$

Then, from recordings of $\mathbf{s}(t)$, the elements a_{ij} of **A** can be found:

$$a_{11} = \tfrac{1}{2}\{[\operatorname{Re}(s_1(t_0)) - \operatorname{Im}(s_1(t_1))] + i[\operatorname{Im}(s_1(t_0)) + \operatorname{Re}(s_1(t_1))]\}$$
$$a_{12} = \tfrac{1}{2}\{[\operatorname{Re}(s_1(t_0)) + \operatorname{Im}(s_1(t_1))] + i[\operatorname{Im}(s_1(t_0)) - \operatorname{Re}(s_1(t_1))]\}$$
$$a_{21} = \tfrac{1}{2}\{[\operatorname{Re}(s_2(t_0)) - \operatorname{Im}(s_2(t_1))] + i[\operatorname{Im}(s_2(t_0)) + \operatorname{Re}(s_2(t_1))]\}$$
$$a_{22} = \tfrac{1}{2}\{[\operatorname{Re}(s_2(t_0)) + \operatorname{Im}(s_2(t_1))] + i[\operatorname{Im}(s_2(t_0)) - \operatorname{Re}(s_2(t_1))]\},$$

where t_0 is arbitrarily chosen so that $\sin \Omega t_0 = 0$ and $t_1 = t_0 + \pi/2\Omega$. Selection of t_0 establishes a reference orientation of the calibration dipole which must be accounted for when making absolute position angle measurements.

The position angle correction for ionospheric Faraday rotation can be obtained from (Davies, 1965)

$$\phi \approx \frac{2.97 \times 10^{-2}}{f^2} \int_Z N B_0 \cos \theta \, dz \quad \text{(MKS units)}$$

where N is electron density, θ is the angle between the magnetic field B_0 and the direction of propagation, and z is the path length. $\int N \, dz$ can be obtained from ionosonde profiles, or the complete integral can be obtained from stationary satellite beacon Faraday rotation measurements. In both cases a correction for the difference in direction between the pulsar signal and the path z must be made.

B. POLARIZATION OBSERVATIONS AND DISPLAYS

A few pulsars, PSR 0833−45, for example, are nearly completely linearly polarized with a smooth linear variation of position angle with longitude. For these pulsars the average polarization profile accurately represents the individual pulse polarization. For other pulsars, however, there are both random and systematic variations of both linear and circular polarization from one pulse to the next, causing apparent depolarization of the average profiles. Since the individual pulse behavior is most closely related to the properties of the radiation mechanism, it is important that individual pulse polarization be studied.

Display of the Stokes parameters of pulsars requires considerable ingenuity, since there are four quantities to be shown as a function of time for each pulse. Direct oscillographs of the four Stokes parameters were made by Clark and Smith (1969), but they are somewhat hard to interpret. Taylor et al. (1971) produced a system to display the linear component using the length of an arrow to represent the magnitude and its orientation to represent the position angle. Taylor (1973) has extended this display to include elliptical polarization, an example of which is shown in Fig. 9. At each increment corresponding to the time resolution is plotted an ellipse whose size, eccentricity, and orientation characterize the polarization. A filled or open ellipse distinguishes between the two hands of the circular components.

Rankin et al. (1974) have used color to display the total intensity, fractional linear $(U^2 + Q^2)^{\frac{1}{2}}/I$, the linear position angle $\frac{1}{2} \tan^{-1} (U/Q)$, and the fractional circular polarization V/I. A distinct advantage of their system is compactness—the polarization state of a sequence of several hundred pulses can be plotted on a single page. Furthermore, systematic polarization features which are preserved or drift in longitude from pulse to pulse are easily recognized as bands of color.

A complete statistical analysis of polarization fluctuations will require application of the techniques described in Section VI, but this has not yet been accomplished.

VI. Intensity Variations with Time

A. Intensity Variations and Their Statistics

In this section we will discuss variations of pulse intensity I. However, it should be realized that in the majority of observations a single antenna is used and the output is often regarded as representing the total intensity. Clearly, the rapidly changing polarization will contribute to the fluctuations and give an inaccurate representation of I. In spite of this remark, it seems likely that results of the statistical analysis discussed in the following sections would not be radically different in true measurements of I.

We are concerned with a time series of pulsar intensities, which it is convenient to represent in the format $I_q(t)$, the intensity from the qth period at intrapulse time t. The intrapulse time is normal time, modulo the period P_1, and will also be referred to by the equivalent longitude $\phi = 2\pi t/P_1$ of the rotating pulsar. It should be stressed that this does not imply that emission received at a certain longitude ϕ all originated at the same geographic longitude on the pulsar, since the emission depends on the beamwidth and direction at each point in the pulsar's

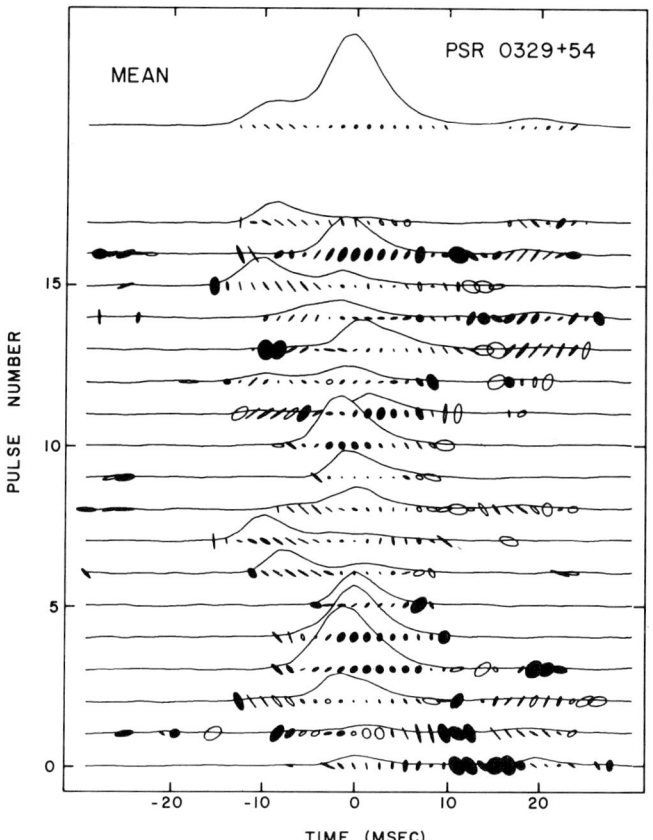

Fig. 9. Complete polarization characteristics of individual pulses from PSR 0329+54. Pulse profiles represent total intensity. Major axes of the ellipses have lengths proportional to total percentage polarization, and the shape of the ellipse is that of the conventionally defined polarization ellipse. If the circular component of polarization is positive, the ellipse is filled in. The longest ellipses shown represent 100% polarization (Taylor, 1973).

magnetosphere. In discussing $I_q(t)$ it should also be noted that the intensity is discretely sampled in t as well as in q (see Section IV), though we will find it convenient to use the equivalent continuous function of t.

The pulse intensities are only nonzero outside a particular window in longitude, as defined by the width of the average pulse profile $a(t)$.

$$a(t) = \frac{1}{M} \sum_{q=1}^{M} I_q(t). \tag{31}$$

This window is typically only 3% of the period so that intensity variation data need only be recorded over a window, say, 10% of the period centered on the mean profile, a saving of 10 to 1 in data handling. Typical observations might record a window of 50 msec with a time resolution of 1 msec from 2000 pulses, giving a two-dimensional array of 10^5 data points. Detailed studies of the variations have been concentrated at frequencies between about 60 and 600 MHz where the signal-to-noise ratio is best. Bandwidths (B) from 50 kHz to 40 MHz and resolutions τ_r from 5 μsec to 100 msec have been used.

The intensities are contaminated by the receiver system noise. This adds a mean power level dependent on the effective system noise temperature, with a fractional rms variation given approximately by $(B\tau)^{-\frac{1}{2}}$. Before processing it is customary to subtract the mean noise level, estimated by the average intensity level outside the pulse window (off-pulse). This can be learned from each period or from the average profile. Figure 10 illustrates the procedure. In what follows we will assume that this subtraction has been done. Receiver noise now only adds a zero-mean white noise of rms σ_N, which can also be determined from the off-pulse intensities by standard variance analysis. The signal-to-noise ratio is defined with respect to σ_N.

We now give some examples of the wide variety of intensity fluctuation observed from pulsars. Figure 11 from Taylor et al. (1975) shows typical sequences of a few hundred pulses with time resolutions of 1–5 msec. They recorded total intensity using a postdetection dispersion removing technique (Section III) at center frequencies of 147 and 400 MHz. Each horizontal row represents the intensity in one pulse period

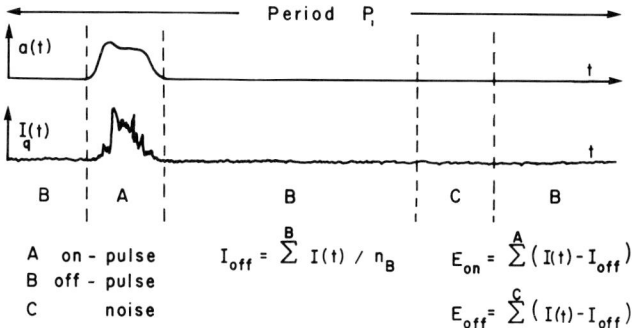

FIG. 10. Estimation of receiver noise correction. The mean off-pulse level I_{off} is estimated from the n samples outside the windows A and C. Then I_{off} is subtracted from the mean of windows A and C to provide estimates of the on-pulse signal and off-pulse receiver fluctuation statistics. One full period is shown for a single pulse intensity record $I_q(t)$ and the average pulse profile $a(t)$.

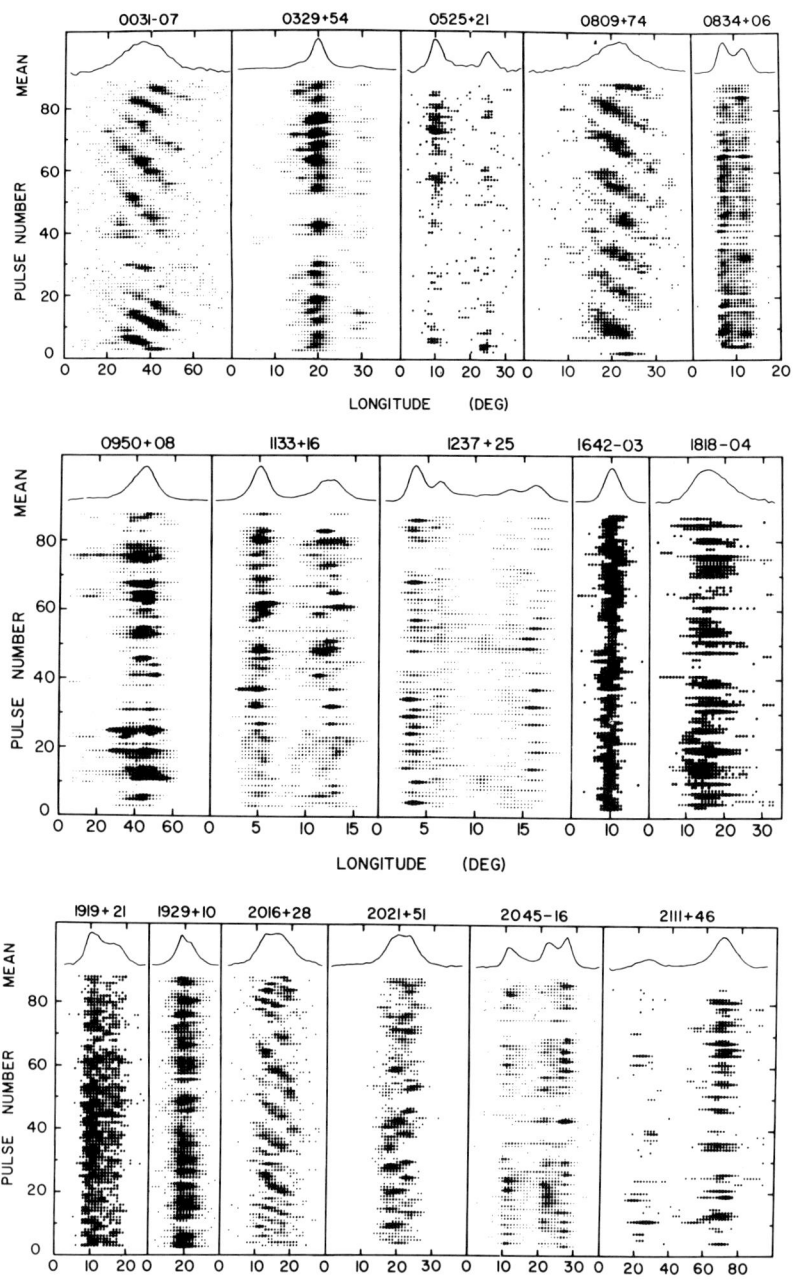

FIG. 11. Longitude–time diagrams and integrated pulse profiles for 15 pulsars observed at frequencies near 400 MHz. Areas of the dots are proportional to received flux density, and time increases upward and to the right (Taylor et al., 1975).

by dots of differing size. The mean profile over 1000 to 5000 pulses, referred to the same pulsar longitude, is also shown. Particular characteristics to be noted are the bands (drifting subpulses), e.g., PSR 0809+74, the contrast between steady pulsars (e.g., PSR 1642—03) and erratic variations (e.g., PSR 0950+08), and the independent variations in differing longitude columns, corresponding to different components in the mean profile (e.g., PSR 1237+25). An example of data taken with a shorter time resolution is shown in Fig. 12 from Hankins (1971). This is a short sequence of pulses from PSR 0950+08 at 111.5 MHz with time resolution of 56 μsec in a 125 kHz bandwidth. Dispersion was removed before detection (Section III) and reveals the extreme variability in strength, shape, and duration of individual pulses.

Before we discuss the processing of many pulses, we should comment on the statistical significance of single-pulse records. For example, which of the various spikes and bumps in Fig. 12 represent significant pulses? To answer this question we need a statistical model of the radiation received from pulsars. A simple model is that the radiation is white Gaussian noise which is amplitude-modulated. We are then interested in studying the pulsed modulating function and not the noise process. The statistics of the pulse intensity at a fixed time t are then χ_n^2 distributed with n degrees of freedom. Here n is the number of independent samples determined from the postdetector response $|K(f)|^2$ and the bandpass power response $|R(f)|^2$, using the definitions of Section IV:

$$\frac{\langle(I - \langle I \rangle)^2\rangle}{\langle I \rangle^2} = \frac{2}{n} = \frac{\int_{-\infty}^{\infty} C_{|R|^2}(f)\,|K(f)|^2\,df}{\int_{-\infty}^{\infty} C_{|R|^2}(f)\,df\,K^2(0)} \quad (32)$$

where the autocovariance function $C_{|R^2|}(f)$ is defined as in Eq. (33) but applies in the frequency domain over an interval much wider than the bandpass. If the postdetector response is much narrower in frequency extent than the bandpass, this reduces to the familiar form given by Bracewell (1962)

$$2/n = 1/\tau_{eq} B_{eq},$$

where τ_{eq} is a time constant defined as the equivalent width of $C_k(t)$, and B_{eq} is the bandwidth defined as the equivalent width of $C_{|R|^2}(f)$ [e.g., $C_k(0)\,\tau_{eq} = \int_{-\infty}^{\infty} C_k(t)\,dt$]. As the resolution is narrowed, the rms fluctuation increases, becoming equal to the mean for no postdetector smoothing; the statistics of I are then exponential. So an answer to the question of significance may be found by comparing the peak deviation in

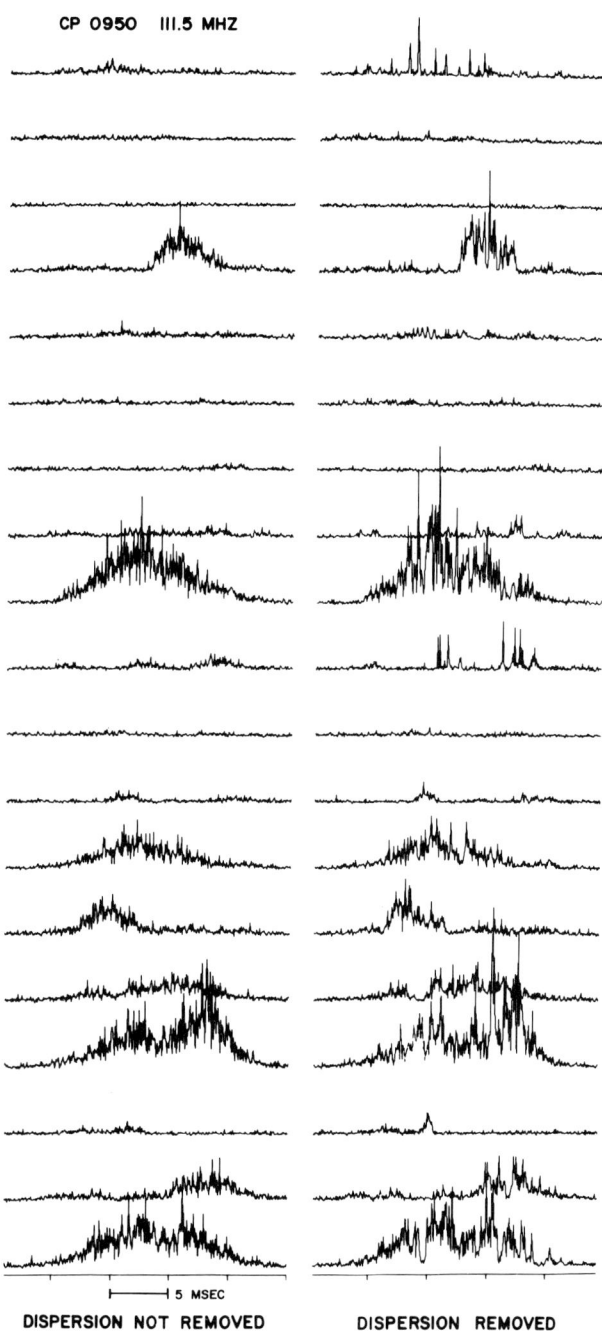

FIG. 12. A small portion of several consecutive pulse periods of PSR 0950+08 is plotted for comparison before dispersion removal and after dispersion removal. The dispersion sweep time across the 125-kHz receiver bandwidth is 2.2 msec, and the postdetection smoothing time is 56 μsec (Hankins, 1971).

question with the expected standard deviation using a local estimate of the mean and Eq. (32). From the χ_n^2 statistics a probability can then be assigned that the observed peak signal represents a modulated increase in the power level. One difficulty is to decide how long an interval defines the local mean. Hankins (1972) computed 95% confidence limits from a running mean over an interval defined by the width of autocorrelation functions (Section VI, B). It may be argued that the model of amplitude-modulated noise has not yet been proved to be accurate. However, if a peak intensity is found significant under the noise model, it is almost certainly significant under any other model.

The examples (Figs. 11 and 12) show both the importance and the difficulty of systematic processing of pulsar signals. We would like to establish the existence of any deterministic processes present in what seem to be random variations. Equally we want to extract the statistics of the random fluctuations. We give in the following sections details of various analysis techniques and what deductions have been made from them.

One question concerns whether the statistics of the variations depend on position t in the profile. This can be tested by modeling the intensity $I_q(t)$ by $a(t)\,j_q(t)$, where $a(t)$ is the average profile and so $j_q(t)$ is the pulse intensity corrected to unity mean (i.e., $\langle j_q(t)\rangle_q = 1$). We should then study how the statistics of $j_q(t)$ depend on t and q. Any answers have to be interpreted in terms of emission from a rotating neutron star. As the star turns we sample the emission on a locus of longitude versus time, and we cannot readily distinguish between temporal variations of intensity and angular variations (due to a narrow beam). One specific test for this is discussed under the heading of drifting subpulses.

B. Correlation Functions for Variations within a Pulse Period

The erratic variations of pulse intensity have to be described statistically, for which purpose the autocorrelation function (acf) provides an obvious tool. We first define a particular version of the autocovariance (acv) from data of finite length T, for any function $g(t)$:

$$C_g(\tau) = \int_0^T g^*(t)g(t+\tau)\,dt. \qquad (33)$$

For sampled data an equivalent sum of lagged products would be computed. For example, the pulse intensity $I_q(t)$ has zero mean outside the window on the profile (after the off-pulse level is subtracted). Thus, $C_I(\tau)$ does not depend on T, providing that T is longer than the profile. For describing the typical duration of individual subpulses, the following functions are useful. $\langle C_I(\tau) \rangle_q$ is the acv averaged over M pulses. The average acf is then

$$\rho_I(\tau) = \langle C_I(\tau) \rangle_q / \langle C_I(0) \rangle_q. \tag{34}$$

The accuracy of the estimates of ρ_I and $\langle C_I \rangle_q$ will depend on the integration time. A crude estimate of the error, $\sigma_{C_I}(\tau)$, in $\langle C_I(\tau) \rangle_q$ can be derived from Eq. (B 6.10) of Blackman and Tukey (1958) if the integration time MT is much longer than the effective width of the pulses, and if their acf decreases mononically with τ:

$$\sigma_{C_I}^2(\tau) / \langle C_I(0) \rangle^2 \approx [1 + \rho_I(2\tau)]/(MT/w_e), \tag{35}$$

where w_e is a typical pulse width specified by the equivalent width of $\langle C_I(\tau) \rangle^2$.

Since $I_q(t)$ is the sum of the zero-mean system noise intensity, $N_q(t)$, added to the signal intensity, $S_q(t)$, we can correct the acv by subtracting the noise acv. Thus,

$$\langle C_S(\tau) \rangle = \langle C_I(\tau) \rangle - \langle C_N(\tau) \rangle, \tag{36}$$

where $I_q(t) = S_q(t) + N_q(t); \langle N_q(t) \rangle = 0$.

In normal situations when $1/B \ll \tau$, the off-pulse acv $\langle C_N(\tau) \rangle = T\sigma_N^2 \rho_k(\tau)$, i.e., a spike of height $T\sigma_N^2$ and of width equal to postdetector time constant. For data in which there is little or no smoothing there are some complications since $\langle C_N(\tau) \rangle$ then involves the convolution of $C_k(t)$ and $|C_r(t)|^2$, and, more importantly, there is an extra term to Eq. (36) in the autocovariances of the signal and noise *voltages*. The details are discussed by Rickett (1975). Leaving aside these complications, the error, σ_{C_N}, in the estimate of $\langle C_N(\tau) \rangle$ is given by a formula similar to Eq. (35) with the off-pulse integration time in place of MT. The net error in $\langle C_S(\tau) \rangle_q$ is then given by $(\sigma_{C_I}^2 + \sigma_{C_N}^2)^{1/2}$; errors in $\rho_S(\tau)$ must be computed by combining errors in $\langle C_S(\tau) \rangle_q$ and $\langle C_S(0) \rangle_q$.

Figure 13 shows examples of autocorrelation functions from several pulsars in a bandwidth of 125 kHz (Hankins and Cordes, 1973). Figure 14 shows schematically several distinct features. The spike (one time-constant wide) represents the noiselike nature of the signals. The narrow feature 0.3–1.2 msec wide characterizes the "microstructure," while the broader feature 5–10 msec wide characterizes the "subpulses." For PSR 2016+28 there is, in addition, a second subpulse feature offset at 10 msec, corresponding to the "second periodicity." The widest feature is imposed by the acf of the mean profile.

Specific models may be developed to bring out the significance of the features noted above. For example, if the signal $S_q(t)$ is equal to the product $a(t) \cdot j_q(t)$, where $a(t)$ is the mean profile and the statistics of

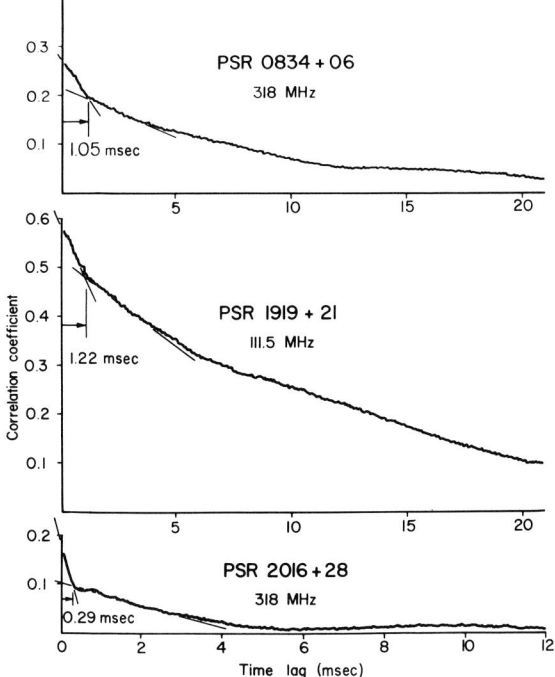

FIG. 13. Intensity autocorrelation functions of several pulsars.

$j_q(t)$ are independent of longitude t, then it follows that the acv of $S_q(t)$ is the product

$$\langle C_S(\tau) \rangle = C_a(\tau) \langle j_q(t) j_q(t + \tau) \rangle. \tag{37}$$

The influence of the mean profile is then apparent and shown in Fig. 14; it depends on the ratio rms divided by mean for the variations $\Delta j_q(t)$.

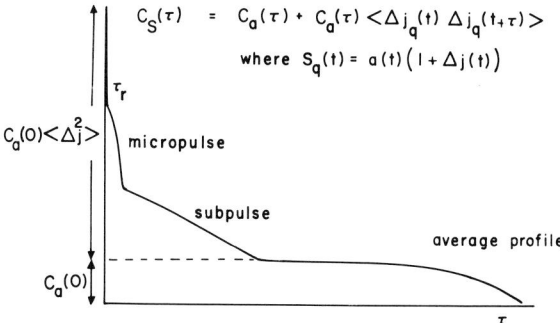

FIG. 14. Diagram of a typical intensity autocorrelation function, showing the receiver resolution time τ_r, the micropulse, subpulse, and average profile features.

In practice, this model is not completely accurate, as is revealed, for example, by Taylor (1974b), in that the observed modulation index defined by

$$\mu^2(t) = \{\langle [I_q(t) - a(t)]^2 \rangle_q - \langle N_q^2(t) \rangle_q\}/a^2(t) \tag{38}$$

was found to vary systematically with time across the pulse—that is, with longitude. For many pulsars he found that the pulses tend to be more deeply modulated near the edges of the mean profile, with modulation indices at or above one.

The noiselike nature of the radiation can be judged by comparing the acf with that expected from a white noise process which is amplitude-modulated by another random function, which could include both subpulse and micropulse variations. The detailed predictions of such a model are described by Rickett (1975) and will not be repeated here.

C. Pulse-to-Pulse Correlation

The presence of correlated structure from one pulse to the next may be revealed by the two-dimensional acv:

$$\langle C_I(r, \tau) \rangle_q = \left\langle \int_{t_1}^{t_1+T} I_q(t) I_{q+r}(t + \tau) \, dt \right\rangle_q.$$

In the previous section we discussed the special case $r = 0$, describing variations during one pulse period. Significant correlation for nonzero values of r indicates subpulses which survive one (or more) periods of rotation; their duration in τ must therefore be related to the width of an angle (or beam) over which the subpulse is emitted. The most startling cases of such correlated structure are provided by the drifting subpulses present in some pulsars (Drake and Craft, 1968a). These give enhanced correlation along diagonal lines in the r,τ domain. The lifetime of a drifting subpulse would be given by the extent in r along the diagonal line. In spite of its appeal, the two-dimensional acf has not been widely used. Examples are given by Taylor *et al.* (1975) in a search for the presence of drifting subpulses in a number of pulsars. Something that is missing in this approach is the longitude dependence. Without the longitude integration, the correlation function becomes a function of t as well as of r,τ, so that drifting subpulses could be resolved by longitude. At this point the computation is very cumbersome and the interpretation versus three variables is too confusing to be helpful. However, it might well be rewarding to compute the two-dimensional acv from the restricted longitude intervals for each component in the pulse profile. Subpulse drifts which are not linear or which do not remain identifiable across the different components might be revealed without relying on the P_3 periodicity as a signature of drifting subpulses (see Section VI, D). Taylor et al. (1975) showed the cross-correlation between each component in the profile, versus period lag r. This is $C_I(r,\tau)$ with a resolution in τ defined by the width of the components.

Backer (1973) suggested that for PSR 2016+28 even the details of the microstructure survive along the subpulse drift path. This would be a very interesting conclusion, since it implies that the short micropulses are caused by a stable and very narrow beam rotating past the observer rather than by a temporal change of emission. Cordes (1975b) has computed the two-dimensional acv over $r = 0, 1, 2$ to search for statistical support for Backer's suggestion; he found no single feature in the acv for $r = 1$ or 2, as narrow as the micropulse feature present at $r = 0$. His results are thus evidence against a beamed origin for the micropulses.

D. PULSE ENERGY FLUCTUATIONS

The fluctuations of the total pulse energy over one to many periods have been studied extensively since the discovery of pulsars. The energy E_q of each pulse is the integral of the intensity over the duration of the mean pulse profile:

$$E_q = \int_0^{T_w} I_q(t)\, dt, \qquad (39)$$

where 0 to T_w defines an on-pulse window. It is understood that the mean intensity level off the pulse has been subtracted. This off-pulse intensity could be found from the average of many pulses, so that ideally it contributes nothing to the statistical uncertainty in E_q. However, receiver drifts could then contribute an apparent pulse variation. A better choice is to average the receiver level from longitudes outside the mean profile, separately for each pulse period. An off-pulse energy sequence is computed from an equivalent length window off-pulse but with the same average receiver level subtraction. Figure 10 illustrates the procedure. The pulse-to-pulse fluctuations have been studied by both autocorrelation and power spectrum analysis of the time sequence E_q. We will concentrate on the power spectra, since most results have been published as power spectra, and considerable interest has been aroused by the periodic fluctuations which they have revealed.

The methods of power spectrum computation have been described by many authors (e.g., Blackman and Tukey, 1958; Jenkins and Watts, 1968). The most common method involves the discrete Fourier transform (using an FFT algorithm) from, say, n blocks of M pulse energies. The squared magnitudes of the Fourier coefficients are averaged together to yield power spectral estimates. Each estimate involves a frequency resolution function $\sin^2(\pi f M P_1)/(\pi^2 f^2 M^2 P_1^2)$ of width $1/(MP_1)$, at intervals of $1/(MP_1)$ from zero to the Nyquist frequency $1/(2P_1)$ (where P_1 is the pulsar period). Each spectral estimate has χ_{2n}^2 statistics with $2n$ degrees of freedom. The zero-frequency interval is not significant because the mean of each block of samples is normally subtracted before the transform. Of course, smoothing in frequency can be used to improve the stability of the estimates at the expense of frequency resolution. The system noise contributes a white component as revealed by the "off-pulse" spectra; this should be smoothed and subtracted from the "on-pulse" spectra to estimate the pulsar signal spectra.

There are many examples of such analysis. Figure 15 shows the power spectrum of PSR 1919+21 from Lovelace and Craft (1968). Note the frequency resolution, number of degrees of freedom, and off-pulse spectrum. Many authors have omitted such information. The main components found in such spectra can be seen in this example: a flat (white) background over and above the noise level; a very-low-frequency component representing variations over tens of minutes mostly caused by interstellar scintillation, merging into a broader "red" component (extending to 0.1 cycles per period); a narrow line component corresponding to a periodic variation at 4.4 P_1. This turns out to be the average time for successive drifting subpulses to pass through a fixed longitude in the mean profile, as first detected for PSR 1919+21 by Drake and Craft (1968a). The symbol P_3 has often been used for the period of such a line feature in the spectrum.

Other pulsars show some or all of these characteristic components in the spectrum, though most attention has been devoted to the periodic components, which include "long period features" at 20–100 P_1 and "short period" features at 2–5 P_1. In addition, pulse nulling (Backer, 1970b) is sometimes observed, in which the pulse energy goes abruptly to zero for one to many periods and then "turns on" again. No systematic behavior of the nulling phenomenon has been found.

E. Fluctuation Spectra versus Longitude

The previous spectrum analysis can be extended to pulse sequences from narrower longitude bins than the whole pulse width. In a sequence of papers, Backer (e.g., 1970a,c) has pursued this method extensively. The data array $I_q(t)$ is averaged over Δt (say, 1 milliperiod) of longitude t, and each sequence versus q for a fixed longitude interval is spec-

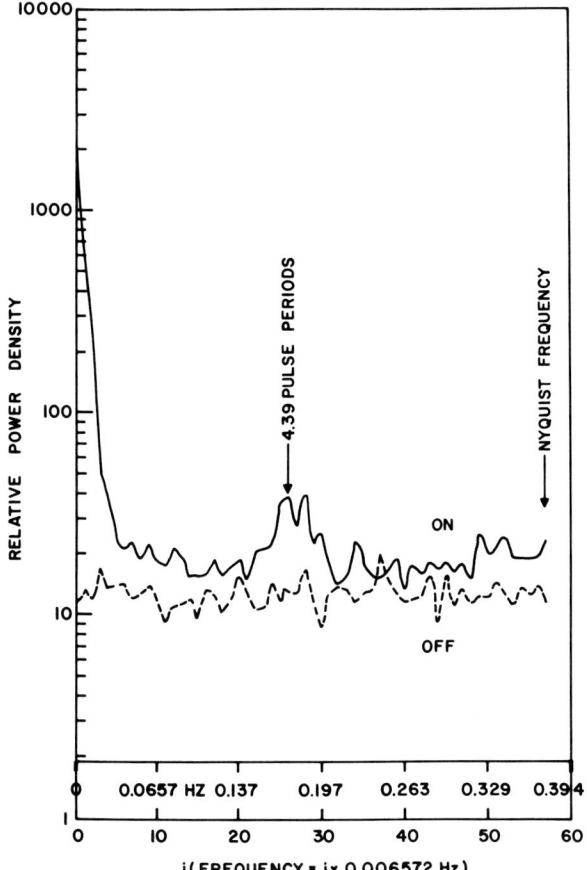

FIG. 15. Power spectrum of the pulse energy variations of PSR 1919+21 at 111.5 MHz (resolution = 0.006572 Hz). Note that the OFF noise spectrum is reasonably flat. 108 degrees of freedom (Lovelace and Craft, 1968).

trum-analyzed as in the previous section. This exaggerates the unavoidable aliasing effect in which fluctuations at a high frequency f, lying between $1/(2P_1)$ and $1/\Delta t$, appear at an aliased frequency between zero and the Nyquist frequency $1/(2P_1)$. From a physical point of view the intensity radiated is a function of pulsar longitude and of time. Ideally we would study the time variations at each longitude, but due to the rotation we only sample each longitude interval once per P_1 the rotation period, and the aliasing problem cannot be avoided.

An example is shown in Fig. 16 from Backer (1973). The spectra for PSR 1237+28 are plotted about every 0.7 milliperiods of longitude and the spectral power is normalized by the square of the mean profile

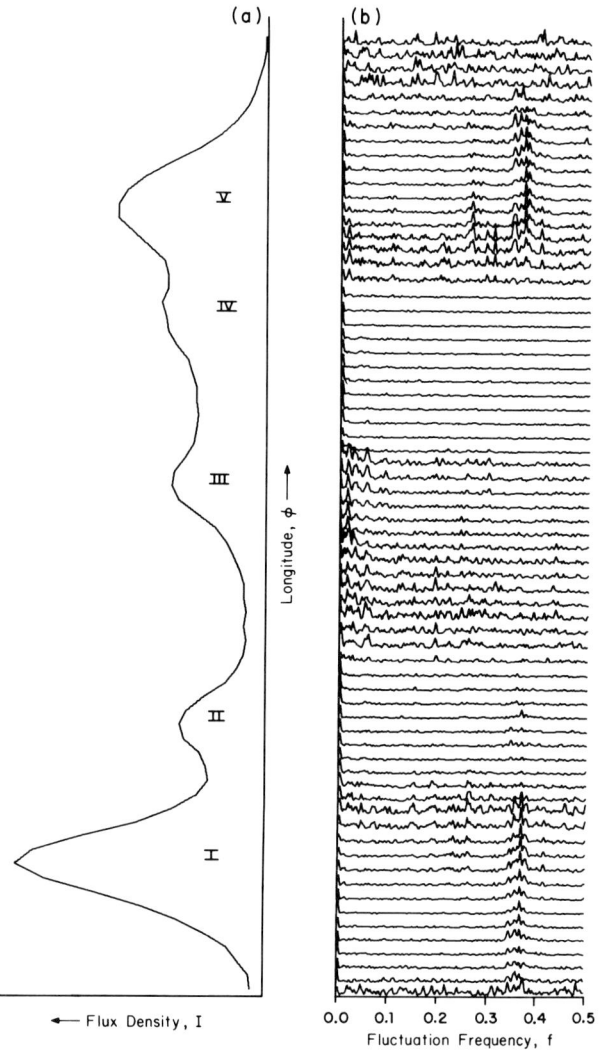

Fig. 16. (a) An average pulse profile of PSR 1237+25 with ordinate of longitude (ϕ) and abscissa of flux density increasing to the left. Five components in the profile are noted. (b) Fluctuation spectra of fixed longitude data with zero of each spectrum opposite the longitude from which the data were taken. Fluctuation power density increases along the ordinate, and fluctuation frequency in cycles per pulse period (c/P_1) along the abscissa. The evaluation of the spectra with longitude shows a mirroring of properties about the midpoint of the pulse (Backer, 1973).

at each longitude. The mean profile itself is plotted to the left, with its five components identified. There is a very complex pattern of behavior, in which the fluctuation spectra are different for the various components but appear to be mirrored about the center of the profile. The short-period feature at 0.37 cycles per P_1 is strong in the outer components, while the long-period feature near 0.02 cycles per P_1 is strong near the center. Similar behavior, though less pronounced, has been seen in other multicomponent pulsars. The spectral estimates from each longitude interval are completely independent, but there is clearly a high degree of correlation between spectral details over several longitude intervals. This suggests that the periodic time variations are correlated over about 4 longitude intervals (i.e., one degree). Cordes (1975a), in analyzing the fluctuation spectra from PSR 1919+21, examined the longitude dependence of various features in the spectra. He found that below 200 MHz the feature at $1/P_3$ and the white spectral power showed a narrow notch at the same longitude as for the average profile. The remaining feature near zero frequency did not show the notch. Such analyses should help pin down the physical origin of the different spectral features.

For the pulsar PSR 1919+21 it is evident (Lovelace and Craft, 1968) that P_3, the short-period fluctuation feature, is the repetition time between successive drifting subpulse bands. This is illustrated in Fig. 17.

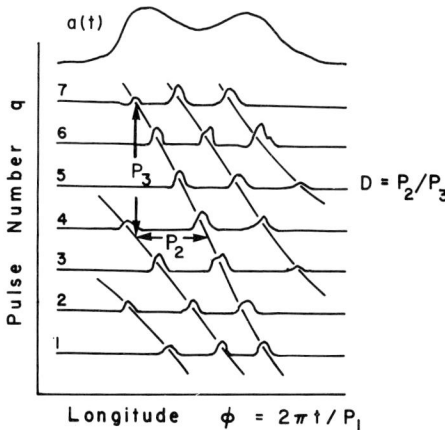

FIG. 17. Description of the drifting subpulse phenomenon. The observer's path is $\phi_q(t)$ which crosses 360° of longitude in one pulse period of time. Time is then measured in pulse number q. The pulsed emission is found in a narrow region of longitude with an average level given by $a(t)$, the average pulse profile. Narrow subpulses are found to drift along paths $\phi_m(t)$, with m indexing successive paths. These are separated, nominally, by P_2 in longitude and P_3 in time (Backer, 1973).

Backer (1973) utilized the phase of the spectral feature at $1/P_3$ to demonstrate this relationship. For a constant drift rate D (seconds per P_1) and well-defined subpulse period P_2 we would find $P_3 = P_2/D$. The phase of the fluctuation spectrum at frequency $1/P_3$ would be $2\pi t/(P_3 D)$ at longitude t, thus giving a way of studying the average drift path. Backer found that D depended on longitude for some pulsars, for example, taking on an "S" shape for PSR 1919+21, corresponding to reverse drifting subpulses for PSR 2303+30, and showing no systematic behavior for PSR 0834+06. He sought to generalize the drifting subpulse phenomenon to include all these cases as well as those of PSR 0031—07 and PSR 0809+74, which show clear drifting bands [in plots of $I_q(t)$], but for which P_3 varies with time in a repetitive manner. This is not altogether satisfactory, since only if a clear spectral feature exists at $1/P_3$ can its phase be used to reveal the average drift path. Another difficult case is PSR 2016+28, in which the drifting bands are very pronounced, though only weakly periodic; there is no spectral line since there appear to be random variations of drift rate and phase; consequently, P_3 is very poorly determined. It has been suggested (Taylor 1974b) that for this pulsar (and also for PSR 0031—07) P_3 and D vary such that P_2 remains constant. The upshot of this discussion is that it is not clear which of P_3 or P_2 or D defines the *basic* character of the drifting subpulse phenomenon. Backer's generalized drifting subpulse idea stressed P_3 and is the most well-developed approach to the overall problem of intensity variations. However, other interpretations may also be viable.

F. Intensity Distribution Functions

Pulse intensity variations have a large random component and so the probability distribution function (pdf) for intensity adds new information to the power spectrum description, which was discussed in previous sections. The pdf of pulse energies E_q is readily computed and shows a variety of forms among the various pulsars (e.g., Lang, 1971). However, the influence of interstellar scintillations (with their own pdf) must be removed to reveal the intrinsic pdf. Backer (1971) corrected the intensities by dividing by an estimate of the interstellar scintillation strength. As discussed in Section VIII, such an estimate is obtained from the average over about 100 periods of the pulse intensity. This is possible since the short-term intrinsic fluctuations cover time scales of up to a few periods, while the interstellar scintillations cause fading with typical time scales of 2 to 20 minutes. The correction is a division since the scintillation modulates the emitted intensity. Scintillation causes the intensity to approach zero on occasions, and the data should then be rejected since division by a small number amplifies the uncertainties. This editing should not distort the pdf since there is no correlation between the intrinsic variations and the minima of interstellar scintillation.

The effect of system noise is to convolve the signal pdf with the noise pdf of rms width σ_N. In principle the resultant could be deconvolved, but it is simpler to convolve model pdf's with that of the noise before comparison with the observations. Backer (1971) found reasonable fits between his observed pdf's and χ^2 distributions with from one to ten degrees of freedom for different pulsars. The most important aspect of the pdf's is their modulation index (the ratio of rms to mean intensity) as defined in Eq. (38); the number of χ^2 degrees of freedom is a convenient way to characterize this. Several authors have analyzed the intensities in fixed longitude intervals and plotted the modulation index against longitude. As noted previously, $\mu(t)$ is greater than one and largest near the edges of the profile for many pulsars. Hence, the mean profile should perhaps be regarded as a distribution function for the longitudes of arriving subpulses rather than for their intensities. At the edge of the profile the subpulses are likely to be just as strong as at the middle but with a greater standard deviation—that is, less frequent in occurrence.

G. Estimation Error in Mean Profiles

The stability of the estimates of the mean profile depends on the pulse-to-pulse intensity variations. We can express the rms variation in an estimate obtained from averaging N pulse periods in terms of the modulation index $\mu(t)$. The fractional error is $\mu(t)/\sqrt{N_1}$, where $N_1 (\leq N)$ is the number of independent samples of variation from pulse to pulse at fixed longitude. For most pulsars there is little pulse-to-pulse correlation and so $N_1 = N$, but for longitudes which are dominated by the P_3 variations, $N_1 \sim NP_1/P_3$. The errors are correlated over longitude intervals characterized by the width of the acf-versus-t, as discussed in Section VI, B. With these points in mind, the estimation of a mean pulse profile $\hat{a}(t)$ should be accompanied by an estimate of the modulation index and profile error $\sigma_a(t)$ in the following fashion:

$$\hat{a}_{\text{on}}(t) = \frac{1}{N} \sum_{q=1}^{N} I_q(t); \quad \hat{a}_{\text{off}} = \frac{1}{N} \sum_{q=}^{N} I_{\text{off}q};$$

$$v_{\text{on}}(t) = \frac{1}{N} \sum_{q=1}^{N} I_q^2(t) - \hat{a}_{\text{on}}^2(t); \quad (40)$$

$$v(t) = v_{\text{on}}(t) - v_{\text{off}}$$
$$\mu^2(t) = v(t)/(\hat{a}_{\text{on}}(t) - \hat{a}_{\text{off}})^2$$
$$\sigma_a^2(t) = v_{\text{on}}(t)/N_1$$

VII. Intensity Variations with Frequency

A. Frequency Dependence of Pulsar Radiation

We discuss the general dependence of the pulses on radio frequency, progressing in order of decreasing time average.

1. Average Spectra

One of the least well-determined aspects of pulsar radiation is their average spectra. This is because it requires very long integration times to average out the known causes of time variation, and even then there appear to be intrinsic long-term variations on the order of months, which may or may not involve changes of spectrum.

The basic quantity which is measured is the energy per pulse E in joules/m^2/Hz (as defined in Section VI, D), averaged over a long time. The intrinsic pulse-to-pulse variations are effectively averaged by, say, 10 min integration. However, this is just the time scale of typical interstellar scintillation; the pulse energy must be averaged over many of the characteristic cells in the frequency–time plane (Section VIII). At high radio frequencies above, say, 400 MHz, this requires longer and longer integration, since the scale of the cells increases with radio frequency in both time and bandwidth.

Comella (1972) has made a thorough study of the spectra of 13 pulsars from Arecibo. He discussed in detail the stability of his estimates, and concludes that the pulsar spectra typically maximize in the region 100 to 400 MHz, falling with a power law ($\sim f^{-2\pm 1}$) up to, say, 3 GHz, where there is often a further steepening of the spectrum. The turnover at low frequencies seems real for several pulsars, but the estimates become unreliable due to worsening signal-to-noise ratio below 100 MHz. For the Crab pulsar the pulse broadening due to interstellar scattering causes the low-frequency turnover of the spectrum. This occurs when the pulses are so broadened that they are longer than the period P_1. The Crab pulsar observations of the smeared-out pulsar (the "compact source") and the scattering laws indicate that the intrinsic spectrum keeps on increasing as frequency^{-3} down to about 10 MHz.

2. Pulse Profile versus Frequency

The profiles of the pulsars are remarkably independent of frequency. For example, the main pulse and interpulse of the Crab pulsar are evident from 100 MHz all the way to X-ray frequencies. However, some profiles, especially the double ones, tend to become narrower with increasing radio frequency. No special processing techniques are involved—it is merely found that the width of the components and their separation vary approximately as frequency$^{-\eta}$, where values of η from 0 to 0.3 are found (with even one case, PSR 2016+28, going the other way, pulses widening with frequency). The other type of variation which is found is that the relative strength of the components of a profile changes with frequency. This is most simply stated by saying that each component has its own radio spectrum. No entirely successful model for the radio emission from pulsars has yet been found, but the relative stability of the profiles against frequency provides a very strong constraint on the models.

3. Subpulse and Micropulse Spectra

Direct comparison of pulses recorded simultaneously at two well-separated frequencies shows some similarities in the individual subpulse details when they are aligned according to the dispersion delay. The random nature of the subpulse and micropulse variations requires that the comparison be done statistically. The cross-correlation of intensity records at the two frequencies provides the obvious tool. We write $I_{1q}(t)$ as the intensity at frequency f_1 for the qth emitted pulse and define

$$I_{1q}(t) = a_1(t)[1 + \Delta j_{1q}(t)], \tag{41}$$

where $a_1(t) = \langle I_{1q}(t) \rangle_q$, the average profile at f_1, and $\Delta j_{1q}(t)$ is the subpulse–micropulse variation such that $\langle \Delta j \rangle = 0$. Similar definitions apply for frequency f_2. In order to remove the cross-correlation associated with $a_1(t)$ and $a_2(t)$, we compute the following:

$$\begin{aligned}\int \langle I_{1q}(t) I_{2q}(t+\tau) \rangle_q \, dt &- \int a_1(t) a_2(t+\tau) \, dt \\ &= \int a_1(t) a_2(t+\tau) \langle \Delta j_{1q}(t) \, \Delta j_{2q}(t+\tau) \rangle_q \, dt.\end{aligned} \tag{42}$$

If $<\Delta j_1 \, \Delta j_2>$ is independent of longitude t, then the result can be corrected further by dividing by the profile cross-correlation. However, it is not known *a priori* that the statistic of Δj are stationary over t. In order to provide a normalized cross-correlation function (ccf), the function in Eq. (42) should be divided by the square root of the product of similar acv's at f_1 and f_2 (each corrected for receiver noise as in Section VI, B).

Figure 8 shows the cross correlation [Eq. (42)] from PSR 0950+08 at 111 and 318 MHz (Rickett, 1974). The surprising result here is that the micropulses characterized by the 200-μsec feature of the acv's show strong cross-correlation over 200 MHz. The peak occurs at a time lag slightly different from zero, indicating that the assumed dispersion delay was not quite accurate. The application of this technique in accurate dispersion measurements was discussed in Section III.

Remarkably, few such cross-correlation analyses have been published, though the wideband nature of the subpulses has been recognized by many observers (e.g., Lyne and Rickett, 1968; Craft, 1970). Boriakoff (1973) computed cross-correlations over a 10 MHz separation and also found a high degree of correlation for some pulsars. There are, of course, many processes which reduce the correlation—for example, the frequency-dependent interstellar scintillation and the use of antennas responding to different polarization at the two frequencies. These should reduce the level of cross-correlation but should probably leave the shape of the ccf substantially unchanged.

B. Frequency and Time Resolution

We know discuss spectra measured over narrow frequency intervals. The measure of intensity in conventional radio astronomy is the flux density S in joules per square meter per second per hertz. Therefore, the energy received in time t to $t + \Delta t$ over a bandwidth f to $f + \Delta f$ is $S(t, f) \Delta t \Delta f$ J/m². Thus, flux density seems to be defined as energy per square meter, differentiated with respect to both time and frequency. Of course, such a definition would violate the uncertainty principle that must link functions of frequency and time. Bracewell (1962) was careful to avoid such a definition by specifying that S is the *average* energy flux per second per hertz. However, for pulsars we are concerned with variations with both time and frequency from individual pulses, and so we are always constrained by the reciprocal relationship between the minimum time and frequency resolutions: $\Delta t \Delta f = $ constant (the constant depends on the resolution criterion but is of order one).

If data are recorded from a bandwidth B for a time T (longer than $1/B$), then there are $2BT$ independent sample values. In principle, they can be analyzed as either a time series of $2BT$ samples with minimum time resolution $1/B$ or $2BT$ points in a spectrum with minimum frequency resolution $1/T$ or any intermediate grid of, say, r cells with time resolution $\Delta t = T/r$ by $2BT/r$ cells with frequency resolution $\Delta f = r/T$. Practical methods for spectral analysis (see Ball, this volume, p. 177) include multichannel filters of (fixed) bandwidth, digital real-time autocorrelation analysis evaluated at intervals Δt (Fourier transformed to generate spectra of minimum resolution $\Delta f \sim 1/\Delta t$), on-line (off-line) computation from data sampled before detection, allowing flexible resolution cells.

Evidently, there is also the usual tradeoff between stability of the spectral estimates and the required resolution in frequency and time.

C. NARROWBAND SPECTRAL VARIATIONS

1. Narrowband Spectra

Individual pulses analyzed with fine frequency resolution show a deeply modulated spectrum as, for example, is shown in Fig. 18 for PSR 0950+08. Displayed here is the power spectrum $|V(f)|^2$ of subpulses, plotted successively throughout one pulse. The estimation error in this plot is large since there is only one independent data point in each resolution cell. If the statistics were stationary white noise, the fractional rms error would be unity. The question of the pulsar signal statistics is important in judging the significance of these deep spectral variations.

In order to characterize the spectral variations we once again resort to the acf as a possible description. Figure 19 illustrates the relationships between the various signal quantities. Procedures for computing estimates of the acf, $R_{|V|^2}(\nu)$, their correction for receiver noise and their stability, follow very closely the outline given in Section VI, B and will not be repeated here. Boriakoff (1973) presented some acf computations of pulsar spectra.

2. The Nature of Spectral Variations

The broadband nature of the subpulses (say, 100 MHz) and their short time duration w (about 1 msec, say), together imply that there must be variations in the amplitude $A(f)$ and/or the phase $\phi(f)$ of the complex spectrum $V(f)$. If this were not the case, the pulse width would, of course, be the reciprocal of the total bandwidth (i.e., about 10 nsec). Figure 18 shows the deep modulations of the power spectrum $A^2(f)$; it is clear that the wiggles have widths of the same order as the reciprocal of the subpulse durations. The question thus concerns the nature of those wiggles: are they random across frequency, or perhaps part of a complex dispersion law caused by emission and propagation in the dense relativistic plasma expected in a pulsar magnetosphere? Manchester et al. (1973) report subpulse widths which are independent of frequency, requiring that any dispersion law be surprisingly independent of frequency.

As a first step, the observations can be compared with the predictions of the amplitude-modulated noise model. The variations in $A(f)$ and $\phi(f)$ should then be statistically stationary over frequency and have a typical width of $1/w$. It can be shown (Rickett, 1975) that the acv of the power

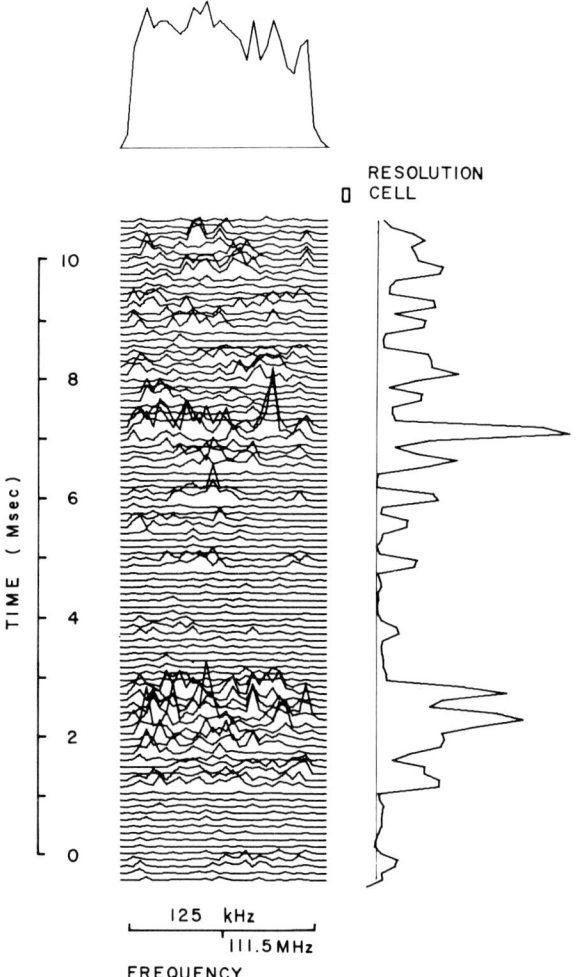

Fig. 18. Narrowband spectra of individual subpulses. Each point of the intensity $I_q(t)$ plotted on the right is the sum of the distribution of intensities across the receiver bandwidth shown in the center. At the top is plotted the spectrum averaged over the pulse. In the limit of many thousands of pulses this would show the receiver bandpass shape (Rickett and Hankins, 1973).

spectrum $R_{|V|^2}(\nu)$ is just the Fourier transform of the acv of intensity variations $R_I(\tau)$, according to the noise model. This gives a quantitative form to the intuitive notion that short pulses will have wide spectral features and vice-versa. It also provides a test for the model. The matter is at present open and must await further analysis.

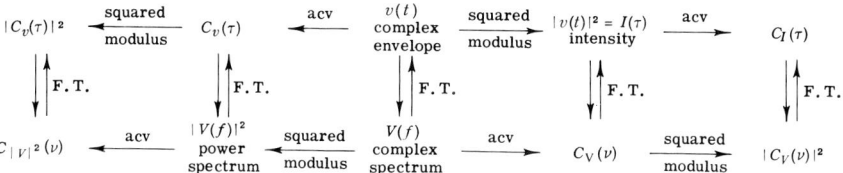

Fig. 19. Relationships between various functions of a pulsar signal. F.T. represents a Fourier transform; acv represents the autocovariance.

VIII. Interstellar Scattering and Scintillation

A. Fluctuations Over Time and Frequency

As radio waves longer than about 10 cm in wavelength propagate through the Galaxy, they are scattered by inhomogeneities in the electron distribution of the interstellar medium. For a point source this causes a broadening of the apparent source brightness distribution, and also scintillation of the intensity observed at the Earth. Most radio sources have a wide enough angular extent that the scintillations are smeared out and the observed intensity is steady. However, the pulsars approximate point sources much more closely and the interstellar scintillation is observed as a deep modulation over both time and frequency imposed on the pulsed intensity. It is convenient to represent the interstellar scintillation by a multiplication of the emitted complex spectrum by $U_q(f)$, for the qth pulse period. In fact, the irregular interstellar propagation is a more complicated process, but $U_q(f)$ can accurately represent the net fluctuations if the emitted spectrum is statistically stationary over the bandwidth used in observing (see Section VII, C). The interstellar scintillations over time and frequency are displayed in Fig. 20 (Rickett, 1970). The pulse signals were spectrum analyzed by gating a digital autocorrelation spectrometer so that it only accepted signals during the on-pulse phase of each period. The spectra averaged over 50 pulses are plotted successively. The frequency resolution is 60 kHz and the net on-pulse integration time was about 1.5 sec, giving an estimation error in each resolution cell of less than 1%. Plotted in Fig. 21 is the autocorrelation of these data against time and frequency, revealing characteristic widths (to $1/e$) of 5 min and 125 kHz. The acf is symmetrical about the frequency axis, indicating that there is no systematic drifting of the scintillation features in the f/t plane. Similar observations (e.g., Ewing et al., 1970) have, however, revealed spectral bands drifting diagonally for a few pulsars. Boriakoff (1973) also made such spectral observations, but seldom found drifting spectral bands.

The data in Fig. 20 are adequately resolved in both frequency and

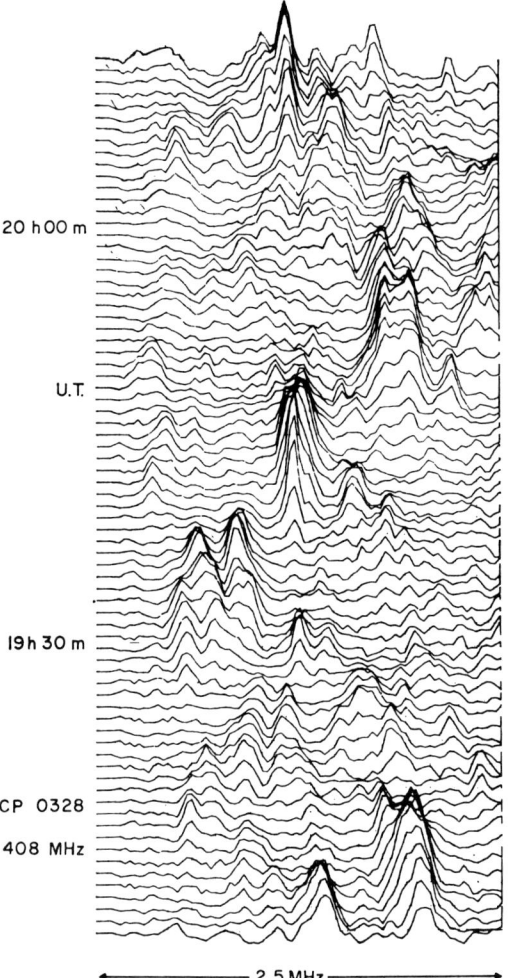

Fig. 20. Spectra from PSR 0329+54 at 408 MHz integrated over about 70 pulses, plotted at successive 50 sec intervals. The frequency resolution is about 60 kHz and the spectra include the receiver bandpass, which gives a gradual cutoff at the edges of the diagram (Rickett, 1970).

time, but that may not always be possible, and we now study how underresolved data should be corrected to reveal true acf widths. The pulse energy in a frequency band centered on f_1 is given as follows:

$$E_{1q} = \int_0^T I_q(t) \, dt, \qquad I_q(t) = \tfrac{1}{2}|y_q(t)|^2,$$
$$y_q(t) = \int_{-\infty}^{\infty} V_q(f) U_q(f) R(f - f_1) \exp{(i2\pi ft)} \, df, \tag{43}$$

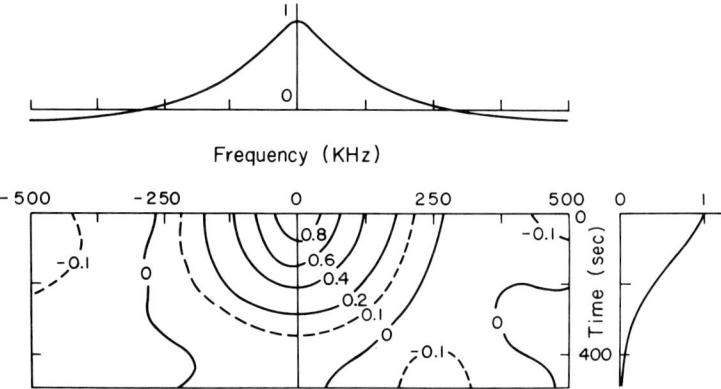

FIG. 21. Contour diagram of the two-dimensional autocorrelation function (in time and frequency) of the data of Fig. 20. Sections along the time and frequency axes are shown (Rickett, 1970).

where $V_q(f)$ represents the (complex) intrinsic pulse spectrum, $U_q(f)$ is the interstellar scintillation, and $R(f - f_1)$ is the bandpass centered on f_1. The observed quantity is $\hat{E}_{f_1 q_1}$, which is the average of E_{1q} from a bandpass centered on f_1 over a time interval T (about 100 pulse periods long) centered on pulse number q_1. Given that the time integration is long enough to average the intrinsic rf spectrum variation, which is incoherent from one period to the next, it follows that

$$\hat{E}_{f_1 q_1} = E_0 \sum^q \int |U_q(f)|^2 |R(f - f_1)|^2 \, df \, \Pi_{(q-q_1),\,T}, \tag{44}$$

where E_0 is the average $\langle |V_q(f)|^2 \rangle$ and $\Pi_{n,T}$ is the rectangular time average window ($\Pi_{n,T} = 1/T$ for $|n| \leq T/2$, $\Pi_{n,T} = 0$ for $|n| < T/2$). Hence, if the statistical variations of $|U_q(f)|^2$ are stationary with respect to q and f, it follows that the autocorrelation of $\hat{E}_{f_1 q_1}$ with respect to frequency and time is the double convolution of the intrinsic acf of $U_q(f)$ with the acf of the bandpass and the acf of the time integration window:

$$\langle\langle E_{q_1 f_1} E_{q_1+r,\,f_1+f} \rangle\rangle = \sum_{q=-\infty}^{\infty} \int df \, \rho_{|U|^2}(f, q) C_{|R|^2}(f + \nu) C_\Pi(q + r), \tag{45}$$

where

$$\rho_{|U|^2}(f, q) = \langle\langle |U_{q_1}(f_1)|^2 |U_{q_1+q}(f_1 + f)|^2 \rangle\rangle,$$
$$C_{|R|^2}(f) = \int |R(f_1)|^2 |R(f_1 + f)|^2 \, df_1,$$
$$C_\Pi(q) = \sum_{q=-\infty}^{\infty} \Pi_{q_1,T} \Pi_{q_1+q,T}.$$

(The $\langle\langle\ \rangle\rangle$ represents the long-term average over many scintillation time scales.) The latter two functions are deterministic and can be accurately estimated so that the intrinsic acf, $\rho_{|U|^2}$, can be recovered by deconvolution. The accuracy of the result is, of course, limited by the combination of a correction for system noise and for intrinsic variations and the estimation error governed by the number of independent data points recorded.

The only report of such processing is by Shitov (1972), who corrected the apparent frequency acf using an analogous method. Boriakoff (1973) also corrected his frequency acf's but used an approximate technique which involved subtracting the cross-correlation of the filter bandpasses.

B. PULSE BROADENING

The modulation of the spectrum by a random function $U_q(f)$ implies that the pulses will also be broadened by convolution with $u_q(t)$, the Fourier transform of $U_q(f)$. If the variations of $U_q(f)$ are random over frequency but are correlated over a characteristic width Δf_{ISS}, then $u_q(t)$ (the temporal smoothing function for the complex amplitude) will have a basic envelope duration of $\tau_{ISS} \sim 1/\Delta f_{ISS}$, which modulates noiselike variations. Averaging the pulse intensity profile gives the average intrinsic pulse shape convolved with a smoothing function $g(t) = \langle\langle |u_q(t)|^2 \rangle\rangle$. This is the Fourier transform of $\langle\langle U_q(t) U_q^*(f + \nu)\rangle\rangle$, the complex correlation versus frequency. For most well-behaved random variables, the fluctuations in phase and amplitude occur over similar characteristic frequency intervals; consequently, the width of this complex acf is the same as the width of the fluctuations in the power spectrum $|U_q(f)|^2$, as characterized by the width of $\rho_{|U|^2}$. Hence, one expects a reciprocal relation between the time duration of the pulse broadening function $g(t)$ and the width of the frequency structure imposed by interstellar scintillation. Salpeter (1969) first pointed to such a relationship. Sutton (1971) expressed it analytically, for the special case of Gaussian shapes which he assumed for all the correlation functions associated with scattering in a screen.

The theory of interstellar scattering is still in a rather inadequate state, with frequent appeal to the "phase screen" model, equivalent in strength to the total phase variations occurring along the propagation path, but assumed to be localized at the midpoint of the total distance z. If this view is taken, there is a close relation between the temporal smearing τ_{ISS} and the average width θ_s of the angular power spectrum of the scattered radiation (Cronyn, 1970):

$$\tau_{ISS} = (z/4)\theta_s^2. \tag{46}$$

This and the generalization, worked out by Williamson (1972) for scattering distributed all along the path, rely on a ray theory. No adequate wave solutions of the pulse broadening have been published. For the special case of a thin phase screen whose phase correlation function takes a Gaussian form, the angular spectrum of scattered waves is also of Gaussian form (Fejer, 1953) and the temporal broadening function becomes a truncated exponential $e^{-t/\tau_{ISS}}$ (Cronyn, 1970). Modeling the extended interstellar scattering by its equivalent phase screen leads to the prediction that $\tau_{ISS} = Sf^{-4}$, where S depends on the average distance and strength of the irregularities. Salpeter (1969) has argued that departures from this model would give rise to even steeper dependence on frequency than the fourth power. The matter is not settled, however, since he did not analyze all possible cases.

On the experimental side, the f^{-4} law has been found compatible with several observations (e.g., Lang, 1971), though in some cases (e.g., Taylor and Huguenin, 1971) a smaller exponent has been observed. The methods of observation and data analysis were not very sophisticated in these treatments. By contrast, Rankin and Counselman, in a series of papers on the Crab pulsar, have approached the problem more thoroughly. Fig-

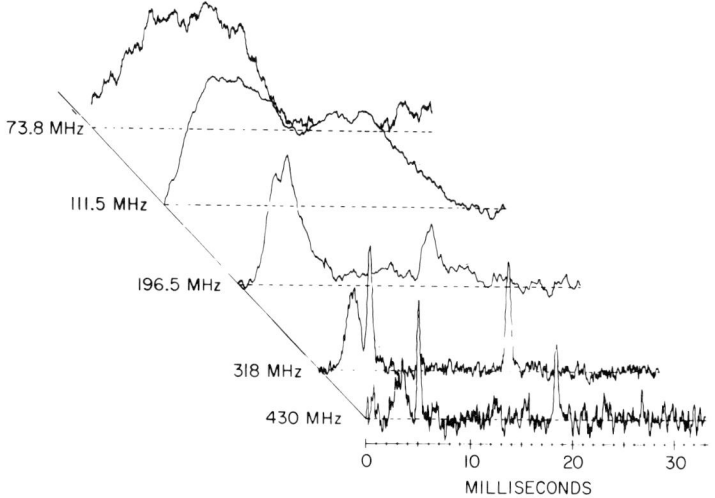

FIG. 22. Average waveforms (over about 2×10^5 periods) observed at 430, 318, 196.5, 111.5, and 73.8 MHz, from Rankin et al. (1970). These observations were not all simultaneous but were made on four days in 1969 between July 30 and August 22. The relative phases shown here are arbitrary. Radiometer time resolutions, in order of decreasing frequency (front to back), were 0.25, 0.42, 0.55, and 11.5 msec.

ure 22 shows an example of their average pulse profiles from PSR 0531+22 at five radio frequencies, clearly demonstrating the frequency-dependent pulse broadening. The problem they tackled was to extract both the dispersion coefficient D [Eq. (11)] and scattering constant S from these measurements, for which the f^{-4} law had already been established (Rankin et al., 1970). We give a short summary of their processing technique.

They computed the Fourier coefficients of the pulse profiles as observed at frequencies of 73.8, 111.5, 196.5, and 430.0 MHz and concentrated on the arrival time information contained in the phases of these coefficients. Conselman and Rankin (1971) analyzed over 100 sets of profile data and computed from each set the phases of the various Fourier coefficients that were significant compared to the estimation errors due to noise and integration time. Since at 196.5 MHz the signal-to-noise ratio was very good, they consistently obtained very accurate phases for the first six Fourier coefficients. Thus they represented the data at the other frequencies by the observed phase of each significant Fourier coefficient relative to that at 196.5 MHz, giving 100 sets of data each comprising 13 phase differences (6 from 430/196.5, 6 from 111.5/196.5, and only 1 from 73.8/196.5 MHz due to lower signal-to-noise ratio at 73.8 MHz). The difference between the 13 observed values and the values predicted from various theoretical models was subject to a least-squared error fitting procedure with respect to the model parameters, inculding the dispersion and scattering constants D and S. From Eq. (30) the theoretical profile includes the convolution of the following:

(a) An intrinsic profile $a(t)$ assumed independent of radio frequency and estimated from the 430 MHz data where both dispersion and scattering broadening were unimportant.

(b) The interstellar scattering function $g(t/Sf^{-4})$, whose shape was also described by one or more free parameters.

(c) The dispersion delay and broadening in the receiver, whose power response $|R(f - f_0)|^2$ is converted into a time function

$$|R[(t - D/f_0^2)f_0^3/(2D)]|^2$$

centered on the arrival time at band center frequency f_0 [see Eq. (23)].

(d) The impulse response of the post-detection filter, denoted by $k(t)$ with time constant τ.

The predicted phase θ_n of each Fourier coefficient was computed from the resultant theoretical profile. The mean-square residual between observed phase difference ϕ_n and θ_n was weighted according to the inverse square of the estimated error σ_n in each observation, thus

$$Q^2 = \sum_{n=1}^{13} \left(\frac{\phi_n - \theta_n}{\sigma_n}\right)^2.$$

If the shape of $g(t/Sf^{-4})$ were known *a priori*, the problem would be a straightforward (nonlinear) estimation by minimizing Q^2 with respect to parameters D and S. Counselman and Rankin tested a variety of shapes by expanding $g(t)$ as a linear series of Laguerre functions, adding the coefficients of the series to the list of parameters to be determined. The Laguerre functions involve the products of powers of t and an exponential decay. Thus, the possible series of the first few Laguerre functions includes, as special cases, $\exp(-t/\tau_{ISS})$ and $[\exp(-\tau/\tau_{ISS})]t/\tau_{ISS}$, functions which had previously been found to be reasonable fits to the observed profiles (e.g., Lang, 1971; Rankin et al., 1970). Counselman and Rankin found, surprisingly, no significant improvement in the fit to a linear sum of the first four Laguerre functions over that for the simple form $[\exp(-t/\tau_{ISS})]t/\tau_{ISS}$. However, the fit to a single truncated exponential was poor, with a minimum of Q^2, three times greater than the optimum. They also attempted fits to $g(t)$ modeled by the convolution of (up to 11) truncated exponentials, each with its own scattering parameters ($\tau_i \equiv S_i f^{-4}$), but found no significant improvement in the fit over that for two exponentials, which reduces to

$$g(t) = \frac{\exp(-t/\tau_1) - \exp(-t/\tau_2)}{\tau_1 - \tau_2}, \quad t \geq 0; \qquad (47)$$
$$g(t) = 0, \qquad t < 0.$$

Indeed, the best-fitting ratio of τ_1/τ_2 was within 5% of unity, for which the above expression degenerates to $[\exp(-t/\tau)]t/\tau$. They suggested that the scattering occurred in two distinct "screens," for instance, the Crab Nebula and the general interstellar medium.

Following this thorough analysis of the scattering function for PSR 0531+21, Rankin and Counselman (1973) reported that measurements covering 1969–1971 revealed the scattering function was changing in shape and duration as also was the dispersion constant D. The changes occurred in such a way that the fit to "two-screen" model remained good, with one screen remaining constant (S_1, say), while large changes occurred in the other—up to twice the original value and then down to nearly zero over the course of the 22 months studied. The interpretation which they put forward was that the interstellar scattering remained constant, but that there were changes in the irregular plasma in the Crab

Nebula itself, causing changes in both the dispersion and scattering. However, Rankin (1974) has since reported preliminary results on subsequent measurements for which the other scattering parameters, S_1, have also changed—which throws some doubt on the two-screen hypothesis. Cronyn (1974) suggested that these variations are all manifestations of irregularities in the interstellar medium. Changes over many months are seen to be caused by movement of very large irregularities across the line of sight. These change the strength and form of the interstellar scattering.

C. MULTIPLE-SITE OBSERVATIONS

1. Angular Broadening

The scattering manifests itself in another way—by broadening the apparent brightness distribution of any radio source observed through the interstellar medium. Observations with an interferometer therefore reveal the effects of the scattering. The output of an interferometer is the (complex) visibility function:

$$\Gamma(\mathbf{r}) = \langle v(\mathbf{r}_1, t) v^*(\mathbf{r}_1 + \mathbf{r}, t) \rangle_t.$$

This represents the amplitude and phase of the interferometer fringes. In practical measurements of the visibility the fringes are averaged over a certain time. Integration times could be as short as a fraction of a second or as long as many minutes. The scattering causes $v(\mathbf{r}_1, t)$, to fluctuate in amplitude and phase, and hence for short integration time there will be fluctuations in the amplitude and phase of $\Gamma(\mathbf{r})$, which Cronyn (1973) has referred to as visibility scintillations. For integration times longer than the time scale of the phase scintillation, the average visibility is reduced, as it would also be for observation in a bandwidth much wider than the scintillation bandwidth Δf_{ISS}.

Once again the prime example comes from the Crab Nebula pulsar PSR 0531+21. Erickson et al. (1972) and Vandenberg et al. (1973) have made low-frequency VLBI observations (74 and 111 MHz) to extract both the angular scattering and pulse broadening parameters simultaneously. Their results showed that the relation Eq. (46) between angular scattering and pulse broadening is not obeyed, and that the irregularities which determine the broadening function are probably distinct from those determining the apparent angular diameter. More observations and analysis along these lines will be necessary to unravel the complicated story which emerges from the Crab pulsar.

2. Interstellar Scintillation

The second type of two-site observation is a comparison of the intensity scintillation at the separation **r**, without maintaining a phase reference between the two observations. The time delay between similar scintillation fades at two (or more) sites allows an estimate of the velocity of the scintillation pattern with respect to the Earth. The cross-correlation function of the two intensity series can be computed in a fashion similar to the autocorrelation [Eq. (34)]. The cross-correlation function in general peaks at a nonzero time lag, which reflects the time taken for the pattern to drift between two sites. If the average drift velocity is V at angle θ to the baseline **r**, then the time lag t_r gives a relation

$$1/V \cos \theta = t_r/r.$$

From more than one baseline these can be combined to give both speed and orientation of the pattern drift motion.

The above rather simplistic approach has been used successfully in pulsar scintillation observations over the baselines of 3000–6000 km, which are necessary to detect a significant time shift, yielding speeds of 50 to 1000 km/sec (Lang and Rickett, 1970; Galt and Lyne, 1972). The latter authors observed a circumpolar pulsar for 24 hours with two telescopes, and obtained a 360° range of baseline orientations as the Earth rotated. Their observation leads to a large velocity for the scintillation pattern of PSR 0329+54, equal to 350 km/sec. The most reasonable explanation seems to be a large proper motion of the pulsar itself, since interstellar velocities are unlikely to be greater than 100 km/sec.

More sophisticated analysis of the pattern drift measurements allows for the possible effect of rearrangement of the pattern as it moves. As discussed originally for ionospheric scintillation, Briggs et al. (1950) pointed out that apparent pattern velocities estimated as above are biased high by pattern rearrangement. They prescribed a method for correcting this effect. Armstrong and Coles (1972) have reviewed the various relevant techniques in the context of interplanetary scintillation. They conclude that a disadvantage of all such methods is that, since they rely on the absolute values of the cross-correlation coefficients, the estimates are very sensitive to errors in normalizing the cross-correlation functions. Of course, the obvious source of decorrelation is the receiver system noise, which in principle can be accurately estimated and corrected for in the same way that acf's are corrected for system noise (Section VI, B). The raw covariance function is normalized by the square root of the product of variances at each site after the noise contribution has been subtracted from each variance. Further sources of decorrelation—such as receiver gain fluctuations, the changing effects of polarization, radio frequency interference, etc.—are harder to correct. Such difficulties are less severe in pulsar observations, since the pulsed nature of the intensities naturally selects against many forms of interference. Ables et al. (1974) explored one of these techniques which, as they point out, is nominally insensitive to the intrinsic pulse fluctuations which will be 100% correlated at each site with zero-time offset. Their technique relies on the time lag at the intersection of the acf and ccf and gives a direct estimate of the velocity component along the baseline. Unfortunately, their method is still critically dependent on the accuracy of the ccf normalization, for which it is hard to judge the errors. The conclusion is that the phenomenon is more complicated than required for the simple analysis, but that the data are seldom good enough for the sophisticated analysis. The problems return, as in much of this article, to noise correction and estimation error. This latter problem is particularly severe in that only a few (5 to 50) fluctuation times are observed in a typical 12-hour observation.

IX. Timing Measurements

The most startling aspect of pulsars found from early investigations was the constancy of their periods. Hewish et al. (1968) argued that, since the period did not show a Doppler shift—to be expected if the source of the pulses were in orbit round a star—the pulses could not come from intelligent beings on another planet (disproving the Little Green Men hypothesis!). The subsequent discovery of the gradual slowing of pulsar periods helped establish the rotating neutron star model. The discontinuities (or glitches) in the slowdown of PSR 0833−45 caused much excitement since a sudden change in the physical properties of a distant neutron star were suddenly accessible. It is clear that timing measurements provide a most valuable insight into the physics of pulsars. The methods of analysis are well described in the literature and will only be given in outline.

A. Time-of-Arrival Observations

In view of the pulse-to-pulse variability, the pulse arrival times are specified by assigning an epoch to an average pulse profile, integrated over 5 min or longer. The profile is formed by summing intensities according to the apparent period either in real time (in a multichannel signal averager or an on-line computer) or later from digitized data. The sample interval, time constant, and the bandwidth/dispersion sweep time must be short enough to provide a net time resolution τ_r better than, say, one one-thousandth of a period. In addition, the sampling times must be referred to an atomic standard of the same or better accuracy.

Two techniques have been used to determine the location of the observed profile within the assumed period. In the first method (e.g., Nelson et al., 1970), the profile is modeled by a function $w(t)$ and a least-squares fit is made between the observed profile and $Aw(t - t_0) + B$, from which the parameters A, B, and t_0 can be determined; t_0 gives the arrival time relative to the assumed period. The alternative method which has been widely used (e.g., Richards et al., 1970), involves the cross-correlation of the observed profile and the model $w(t)$ (often called a template). The location t_0 of the peak correlation is found, for example, by fitting a parabola between the three highest points of the cross-correlation function. Other schemes are also possible, such as the midpoint at one half the maximum correlation, or the first moment, or the median of the ccf. A characteristic arrival time at the observatory is then computed from this pulse location t_0 and the absolute time of a particular period, chosen from the middle of the integration interval.

B. Errors in Arrival Time

The main sources of errors in the arrival times are as follows.

1. Pulse-to-Pulse Fluctuations

Variations in the observed profile are caused by pulse-to-pulse fluctuations and receiver noise, integrated over the finite observing time. In Section VI we discussed the errors in the mean profile estimate $\hat{a}(t)$ from an integration time T; the standard deviation is $\sigma_{\hat{a}}(t) \sim (P_1/T)^{1/2} \hat{a}(t)$, and the variations are correlated over an interval τ_s, characteristic of individual pulse widths. In order to minimize the effect of receiver noise the net time resolution τ_r should be adjusted to match τ_s. The cross-correlation of the template and $\hat{a}(t)$ is the sum of the deterministic correlation of the template with an average profile plus an error term of standard deviation equal to $\sigma_{\hat{a}}(\tau_s/w)^{1/2}$, where w is the width of the average profile. So too w is the characteristic time scale of the error term in the cross-correlation. Hence the approximate standard deviation σ_{t_0} in the arrival times is given by

$$\sigma_{t_0} \sim w(\tau_s/w)^{1/2} \sigma_{\hat{a}}/\hat{a},$$

therefore

$$\sigma_{t_0} \sim (\tau_s w P_1/T)^{1/2}. \tag{48}$$

For a 1 sec pulsar, 20 min integration, profile width $w \sim 20$ msec and subpulse width $\tau_s \sim 1$ msec, the timing error should be about 0.15 msec—that is, less than 1% of the profile width. The presence of system noise can also contribute to timing errors, but in most cases the pulsar signals are much stronger than the system noise averaged over many minutes.

2. Polarization

For data observed with a linearly polarized antenna, the rotation of the polarization angle through the average profile can lead to a systematic distortion of the apparent profile and hence to errors in the arrival times which depend systematically on the position angle of the antenna. Hunt (1971) estimated that a 25% linearly polarized profile, with 180° swing of position angle through the pulse, would cause a time error of about one thirtieth of w. Since the degree of circular polarization in the mean profiles is so small, similar errors made with a circularly polarized antenna are less by a factor of ten and can normally be ignored. The best solution involves the sum of two orthogonal polarizations to yield the total intensity (Stokes parameter I), so eliminating polarization errors (e.g., Manchester and Peters, 1972).

3. Interstellar Scintillation and Dispersion

In order to achieve time resolution τ_r, a bandwidth no wider than $\alpha \tau_r$ must be used, where α is the dispersion sweep rate [Eq. (9)]. In order to compare arrival measurements between frequencies, we must know the passband center and maintain it to an accuracy of better than $\alpha \sigma_t$ for a timing error σ_t (i.e., stabilized to an accuracy of $B\sigma_t/\tau_s$, where the resolution τ_r is set equal to the typical subpulse width τ_s).

Interstellar scintillation modulates the pulses over both time and frequency, and at worst can cause a shift of the effective band center by as much as $B/2$, giving a timing error of $B/(2\alpha)$, or $\tau_r/2$ if the time resolution is dominated by dispersion. This problem is worst when B is about equal to the coherence bandwidth of the interstellar scintillations and the integration time is less than or equal to the scintillation fading time.

4. Mode Changing

Evidently the intrinsic stability of the pulse profile is most important in allowing arrival times to be measured. Several observers have studied what variations do occur in pulse profiles averaged over many successive integrations. The conclusions are that most pulsars do indeed have very stable profiles, but that some pulsars show "mode changes," in which for short intervals of time the profile takes on a different form, but then reverts to its original shape. For example, in PSR 1237+25 Backer (1970d) found that the profile occasionally switched from its normal five-component shape to a three-component profile and then back to five, on a time scale of a hundred or more periods. In PSR 0329+54, Lyne et al. (1971) observed a more subtle effect in which the ratio of the strengths of the outer components changed to anomalous values and then back. Evidently, if unnoticed, the mode changing could be a major contributor to timing errors up to, say, 30% of the profile width w. Minor unrecognized mode changes are a possible source of variability in all measured arrival times at the level of, say, 10% of w. A useful monitor for such effects is the minimum residual mean-squared error between the observed profile and the assumed template, which should therefore be included in the computations whenever possible. (For an example, see Boynton et al., 1972.)

In comparing the importance of all the sources of error it is clear that only under favorable circumstances can arrival times be measured to one one hundredth of the profile width w, as implied in Section IX, B, 1. If any of the other errors are important, accuracies of nearer one tenth of the profile width are to be expected. Many observers have used the observed variation over successive integrations as error estimates; however, this can only be done after the period-fitting process and does not eliminate the polarization problem (Section IX, B,2).

C. Barycentric Arrival Times

Any observatory undergoes a complex motion through space due to the spinning of the Earth, etc., and consequently, measured pulse arrival times must be corrected to a standard inertial point, namely, the center of mass (barycenter) of the solar system. Table I lists the various periodic terms in the motion of a point on the Earth with respect to each term. The effects must be combined to give a net vector $\mathbf{r}$ from the observatory to the barycenter. Fortunately, accurate ephemerides have been compiled for the Earth's motion by radar astronomers from their studies of signals reflected from the moon and planets (e.g., see Hagfors and Evans, 1968). The arrival time for infinite frequency at the barycenter t_B is then calculated from the observed arrival time t_{obs} as follows:

$$t_B = t_{obs} + [(\mathbf{r} \cdot \mathbf{n})/c] - D/f^2 + \theta, \qquad (49)$$

where $\mathbf{n}$ is a unit vector in the (assumed) direction of the pulsar, D is the dispersion constant, f is the Doppler-shifted radio frequency, and θ

TABLE I
MOTION ABOUT THE BARYCENTER (BC)[a]

Source of motion	Amplitude as a light time	Synodic period (from Earth)
Spinning of Earth	21 msec	1 sidereal day
Earth–Moon	16 msec	27 days
Earth–Sun	499.005 sec	1 year
Jupiter–BC	2.6 sec	299 days
Saturn–BC	1.35 sec	378 days
Neptune–BC	0.80 sec	267 days
Uranus–BC	0.42 sec	370 days
Pluto–BC	6.6 msec	366 days
Earth–BC	1.5 msec	365 days
Venus–BC	0.9 msec	584 days
Mars–BC	0.26 msec	780 days
Mercury–BC	0.03 msec	116 days

[a] Adapted from Richards (1972).

is the relativistic correction to a clock on the Earth. Due to changes in the orbital speed v and gravitational potential, the differential clock rate between coordinate time t_∞ at an infinite distance from the Sun and time t_e on the Earth is

$$dt/dt_e = [1 - 2GM/(rc^2)]^{1/2}(1 - v^2/c^2)^{1/2}.$$

If this is integrated over the Earth's orbit to second order in the eccentricity e,

$$\theta = 0^s.001661[\sin L + 0.5e \sin 2L + 0.375e^2(\sin 3L - 3 \sin L)],$$

where L is the mean anomaly of the Earth and θ is the required term correcting the clock to the rate of a clock fixed at 1 AU from the Sun (Clemence and Szebehely, 1967).

D. MODEL FITTING

The set of observed barycentric arrival times t_B are referred to some standard epoch t_A and then the differences $t_i = t_B - t_A$ must be compared to a model involving the period evolution and corrections to the pulsar position. First, an integer pulse number must be assigned to each arrival time. The simplest approach is to divide the time t_i by an estimate of the period. The quotient should be close to the required integer N_i if the period is known to an accuracy considerably better than one part in N_i. The apparent period (Doppler-shifted to the barycenter) gives a first-period estimate, which is successively refined by increasing the length of the data span; as this process continues, longer and longer times between observations are allowable without introducing period ambiguities. This may even be extended to several weeks between data points provided the first frequency derivative, $\dot{\nu}$, is also added, so that ΔN is the number of pulses between observations at t_1 and t_2:

$$\Delta N = (t_2 - t_1)\nu + (t_2 - t_1)^2 \dot{\nu}/2.$$

The next process involves fitting a model to the series of data pairs (t_i, N_i). The methods quoted in the literature are all similar in principle, but differ in detail. A model T_i can be constructed for t_i with N_i as the independent variable:

$$T_i = g(N_i),$$

which can be expanded as

$$T_i = t_0 + P_0 N_i + Q_0 N_i^2 + R_0 N_i^3 + \cdots \quad (50)$$

Alternatively, with t_i as the independent variable, the integer N_i can be modeled by a continuous variable, the pulse phase $\phi_i = \int_{t_A}^{t_B} \nu(t)\, dt$, where $\nu(t)$ is the instantaneous rotation frequency. Expanding ϕ_i as a Taylor series, we obtain

$$\phi_i = \phi_0 + \nu_0 t_i + \dot{\nu}_0 t_i^2/2 + \ddot{\nu}_0 t_i^3/6 + \cdots \quad (51)$$

This approach has been widely used (for example, by Richards et al., 1970; Papoliolios et al., 1970; Manchester and Peters, 1972); least-squared error estimates of the parameters ϕ_0, ν_0, $\dot{\nu}_0$, etc., can be made directly, since the model is linear in the parameters. Similarly, Hunt (1971) and Duthie and Murdin (1971) estimated the parameters t_0, P_0, Q_0, etc.

Since in both Eqs. (50) and (51) the first-order term dominates by many orders of magnitude for intervals up to many years, it is not important which approach is used. Duthie and Murdin (1971) give the correspondence between the two sets of parameters:

$$\begin{aligned} P_0 &= 1/\nu_0 \\ Q_0 &= -\tfrac{1}{2}\dot{\nu}_0/\nu_0^2 = \tfrac{1}{2}P_0 \dot{P}_0 \\ R_0 &= (\tfrac{1}{2}\dot{\nu}_0^2 - \tfrac{1}{6}\nu_0 \ddot{\nu}_0)/\nu_0^5 = \tfrac{1}{6}(\dot{P}_0^2 + P_0 \ddot{P}_0) \\ \nu_0 \ddot{\nu}_0/\dot{\nu}_0^2 &= 2 - (P_0 \ddot{P}_0/\dot{P}_0^2) \end{aligned} \quad (52)$$

In a physical sense $\dot{P}_0$ or $\dot{\nu}_0$ evidently carry information about the age of the pulsar. Ostriker and Gunn (1969) and Goldreich and Julian (1969) both show that, for the magnetic dipole braking mechanism, the age is estimated by $P/(2\dot{P}_0)$, and the magnetic moment of the neutron star is proportional to $(P_0 \dot{P}_0)^{1/2}$. Attempts have been made (e.g., Huguenin et al., 1971) to correlate these two quantities with other properties of pulsars.

In our discussion so far, it was assumed that the pulsar position was known exactly. In practice this is not the case, and positional errors introduce an annual sinusoidal term into the residual between model and observation. If $\mathbf{n}$ is a unit vector in the assumed direction and $\mathbf{n}'$ is in the true direction, then the extra term $(\mathbf{n}' - \mathbf{n})\cdot\mathbf{r}/c$ is added to the arrival times t_i. If the position corrections are small, $\Delta\alpha$ and $\Delta\delta$ (see, for example, Manchester and Peters, 1972), then the refined model to be fitted to the residual phase would be

$$\Delta(\nu_0)t_i + \Delta(\dot{\nu}_0)t_i^2/2 + A(t_i)\,\Delta\alpha + B(t_i)\,\Delta\delta, \qquad (53)$$

where

$$A = \nu_0 r/c \, \cos\delta_E \cos\delta \sin(\alpha - \alpha_E),$$
$$B = \nu_0 r/c \, [\cos\delta_E \sin\delta \cos(\alpha - \alpha_E) - \sin\delta_E \cos\delta],$$

α, δ are assumed source position, and α_E, δ_E are the coordinates of the Earth with respect to the barycenter, having a one-year period. The parameters $\Delta(\nu_0)$ and $\Delta(\dot{\nu}_0)$ must also be included, since there will be the possibility of coupled errors in the slowdown rate and the position corrections. Equation (53) omits cubic terms, since only for two pulsars has this term yielded an improved fit. All four parameters in this model are linear and so lead to direct least-squares estimates. Pulsar positions have been determined from timing data over more than one year and are accurate to about a second of arc for many pulsars.

Manchester et al. (1974) used equations in this form and added the possibility of a proper motion, $\Delta\alpha$ becoming $\Delta\alpha + \mu_\alpha(t_B - t_A)$ and similarly in $\Delta\delta$, in their analysis of five years of timing observations of PSR 1133 + 16. Figure 23 shows the residual and the fitted curve with a growing sinusoid characteristic of a substantial proper motion $\mu_\alpha = 0 \pm 0.10$ sec arc/year and $\mu_\delta = 0.58 \pm 0.22$ sec arc/year for this pulsar. At its estimated distance of 130 pc this implies a very large velocity, $\sim$380 km/sec, and lends support to the idea that pulsars can be ejected at high speed from the center of their parent supernova.

The optical and radio-timing measurements of the Crab pulsar (PSR 0531 + 21) have been analyzed extensively by many authors. The accuracy is better than for other pulsars, the apparent age is shorter, and the pulsar position is known independently to a good accuracy, making the period evolution more accessible. In spite of these advantages, no entirely satisfactory model has been found. Nelson et al. (1970) attempted a fit to the physically interesting model in which the rate of loss of kinetic energy is a power law in frequency $dE/dt = a\nu^n$. The cases $n = 2, 4$, or 6 correspond, respectively, to braking by stellar wind torque, magnetic dipole radiation, and gravitational radiation. If the moment of inertia I

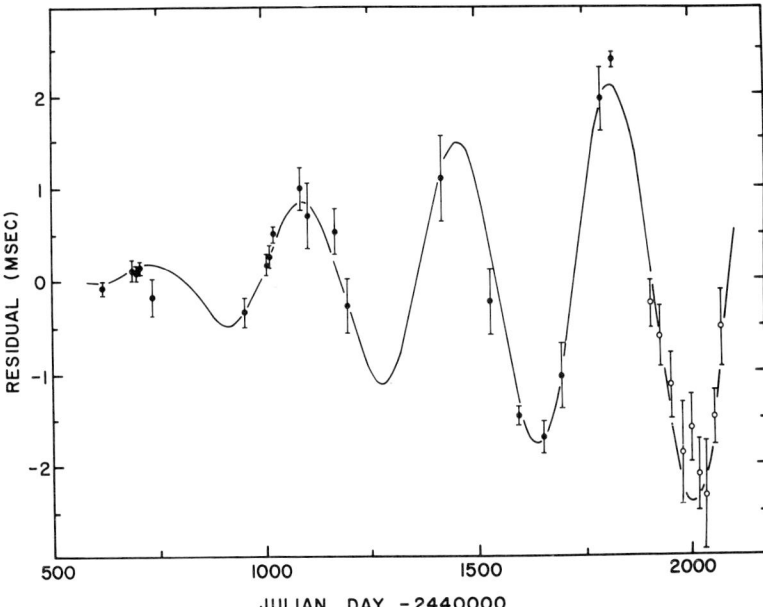

FIG. 23. Residuals for pulse arrival time observations from the National Radio Astronomy Observatory (filled circles) and the Five College Radio Astronomy Observatory (open circles) from a fit of Eq. (53) to the NRAO data only. The plotted points are residuals from the fitted function excluding proper motion terms, and the curve represents the complete function. For the NRAO data, each plotted point is the average of 2 to 10 independent observations; for the FCRAO data, 10 to 20 observations. Error bars are ±2 standard deviations of the mean (Manchester et al., 1974).

remains constant we obtain $\dot{\nu} = (a/4\pi^2 I)\nu^{n-1}$, and hence, by integrating twice, an analytic form for ϕ, which has parameters $4\pi^2 I/a$, n, and ν_0. However, no improvement in the fit was obtained over the power series [Eq. (51)]. The braking index was found to be $n \sim 3.6$, which same value also came from the power series [by $n = 1 + \nu_0 \ddot{\nu}_0/(\dot{\nu}_0)^2$]. They also tried a linear sum of terms with $n = 2$, 4, and 6, but again found no improvement in the fit.

The discovery of a sudden decrease in the period of the Vela pulsar PSR 0833 − 45 (Radhakrishnan and Manchester, 1969; Reichley and Downs, 1969) showed that steady period evolution may not always be a good model. Since then a number of period discontinuities (glitches) have been detected for the Crab pulsar. Studies of the slowdown parameters before and after the glitches have been compared to theoretical predictions for adjustments in the structure of a neutron star (e.g., Baym et al., 1969; Greenstein and Cameron, 1969; Scargle and Pacini, 1971). Boynton *et al.* (1972) have constructed a "glitch function" to be fitted to data containing a suspected glitch. They thus extract four parameters describing each event and the subsequent recovery.

The Crab timing residuals, after obvious glitches and the polynomial evolution have been removed, are still significantly larger than measurement errors (Section IX, B). Boynton *et al.* (1972) concluded that a random process was also present. They compared the statistics of the observed residuals with noise of three forms—a series of random jumps in phase, in frequency, or in rate of change of frequency. By studying the way in which the apparent values of $\dot{\nu}_0$ depended on the total length of the data span, they concluded that frequency noise best represented the data. Suggested physical explanations include matter falling onto the neutron star, or cracking of the solid crust readjusting the surface to an equilibrium shape as the period slowed down.

Recently, Manchester and Taylor (1974) have reported to glitch for the long-period pulsar PSR 1508 + 55, and noiselike variations in the residuals of other pulsars. Clearly this area will continue to be a fruitful source of information on the physics of neutron stars, as longer and longer data spans become available.

Acknowledgments

We gratefully acknowledge the support of the National Science Foundation which, through Grant GP 31686A1, has enabled us to conduct some of the research we have presented here. We wish to thank all of the authors who have provided us with materials and figures for this article, and we are especially indebted to James M. Cordes for his assistance with many parts of this work.

References

Ables, J. G., Slee, O. B., Batchelor, R. A., Krishnamohan, S., Venugopal, V. R., and Swarup, G. (1974). *Mon. Not. Roy. Astron. Soc.* **167**, 31–43.
Armstrong, J. W., and Coles, W. A. (1972). *J. Geophys. Res.* **77**, 4602–4610.
Backer, D. C. (1970a). *Nature (London)* **227**, 692–695.
Backer, D. C. (1970b). *Nature (London)* **228**, 42–43.
Backer, D. C. (1970c). *Nature (London)* **228**, 752–755.
Backer, D. C. (1970d). *Nature (London)* **228**, 1297–1298.
Backer, D. C. (1971). Ph.D. Thesis, NAIC Rep. No. 1. Cornell University, Ithaca, New York.
Backer, D. C. (1973). *Ap. J.* **182**, 245–262.
Backer, D. C., Boriakoff, V., and Manchester, R. N. (1973). *Nature (London), Phys. Sci.* **243**, 77–78.
Baym, G., Pethick, C., Pines, D., and Ruderman, M. (1969). *Nature (London)* **224**, 872–874.
Bergland, G. D. (1969). *IEEE Spectrum* **6**, 41–52.

BLACKMAN, R. B., and TUKEY, J. W. (1958). "The Measurement of Power Spectra." Dover, New York.
BORIAKOFF, V. (1973). "Pulsar Radiofrequency Observations with a Digital Pulsar Processor," NAIC 38. Nat. Astron. Ionos. Cent., Arecibo, Puerto Rico.
BOYNTON, P. E., GROTH, J. E., HUTCHINSON, D. P., NANOS, G. P., JR., PARTRIDGE, R. B., and WILKINSON, D. T. (1972). *Ap. J.* **175**, 217–241.
BRACEWELL, R. N. (1962). *In* "Handbuch der Physik' (S. Flügge, ed.), Vol. 54, pp. 41–129. Springer-Verlag, Berlin and New York.
BRACEWELL, R. N. (1965). "The Fourier Transform and its Applications." McGraw-Hill, New York.
BRIGGS, B. H., PHILLIPS, G. J., and SHINN, D. H. (1950). *Proc. Phys. Soc., London, Sect. B.* **63**, 106–121.
BURNS, W. R., and CLARK, B. G. (1969). *Astron. & Astrophys.* **2**, 280–287.
CAMPBELL, D. B., HEILES, C., and RANKIN, J. M. (1970). *Nature (London)* **225**, 527–528.
CLARK, R. R., and SMITH, F. G. (1969). *Nature (London)* **221**, 724–726.
CLEMENCE, G. M., and SZEBEHELY, V. (1967). *Astron. J.* **72**, 1324–1326.
COLE, T. W. (1972). *Astrophys. Lett.* **12**, 181–183.
COMELLA, J. M. (1972). Ph.D. Thesis, Cornell University, Ithaca, New York.
COOLEY, J. W., and TUKEY, J. W. (1965). *Math. Comput.* **19**, 297–301.
CORDES, J. M. (1975a). *Ap. J.* **195**.
CORDES, J. M. (1975b). (Submitted for publication).
COUNSELMAN, C. C., III, and RANKIN, J. M. (1971). *Ap. J.* **166**, 513–523.
COUNSELMAN, C. C., III, and RANKIN, J. M. (1972). *Ap. J.* **175**, 843–856.
CRAFT, H. D., JR. (1970). "Radio Observations of the Pulse Profiles and Dispersion Measures of Twelve Pulsars," CRSR Rep. No. 395. Center for Radiophysics and Space Research, Cornell University, Ithaca, New York.
CRONYN, W. M. (1970). *Science* **168**, 1453–1455.
CRONYN, W. M. (1973). *Ap. J.* **174**, 181–200.
CRONYN, W. M. (1974). Paper presented at the meeting on "Interstellar Scattering," JPL, Pasadena, California, 1974.
DAVIES, J. G., and LARGE, M. I. (1970). *Mon. Not. Roy. Astron. Soc.* **149**, 301–310.
DAVIES, J. G., HORTON, P. W., LYNE, A. G., RICKETT, B. J., and SMITH, F. G. (1968). *Nature (London)* **217**, 910–912.
DAVIES, K. (1965). "Ionospheric Radio Propagation." US Gov. Printing Office, Washington, D.C.
DRAKE, F. D., and CRAFT, H. D., JR. (1968a). *Nature (London)* **220**, 231–235.
DRAKE, F. D., and CRAFT, H. D., JR. (1968b). *Science* **160**, 758–760.
DUTHIE, J. G., and MURDIN, P. (1971). *Ap. J.* **163**, 1–10.
ERICKSON, W. C., KUIPER, T. B. H., CLARK, T. A., KNOWLES, S. H., and BRODERICK, J. J. (1972). *Ap. J.* **177**, 101–114.
EWING, M. S., BATCHELOR, R. A., FRIEFIELD, R. D., PRICE, R. M., and STAELIN, D. H. (1970). *Ap. J.* **162**, L169–L172.
FEJER, J. A. (1953). *Proc. Roy. Soc., Ser. A* **220**, 455–471.
GALT, J. A., and LYNE, A. G. (1972). *Mon. Not. Roy. Astron. Soc.* **158**, 281–290.
GOLDREICH, P., and JULIAN, W. H. (1969). *Ap. J.* **157**, 869–880.
GREENSTEIN, G. S., and CAMERON, A. G. W. (1969). *Nature (London)* **222**, 862–863.
HAGFORS, T., and EVANS, J. V. (1968). "Radar Astronomy." McGraw-Hill, New York.
HANKINS, T. H. (1971). *Ap. J.* **169**, 487–494.
HANKINS, T. H. (1972). *Ap. J.* **177**, L11–L15.

HANKINS, T. H. (1973). *Ap. J.* **181**, L49–L52.
HANKINS, T. H. (1974a). Paper presented at the Stanford Pulsar Symposium, 1974.
HANKINS, T. H. (1947b). *Astron. & Astrophys., Suppl.* **15**, 363–365.
HANKINS, T. H., and CORDES, J. M. (1973). *Bull. Amer. Astron. Soc.* **5**, 18.
HELSTROM, C. W. (1960). "Statistical Theory of Signal Detection." Pergamon, Oxford.
HEWISH, A., BELL, S. J., PILKINGTON, J. D. H., SCOTT, P. F., and COLLINS, R. A. (1968). *Nature (London)* **217**, 709–713.
HUGUENIN, G. R., and TAYLOR, J. H. (1969). *Ap. Lett.* **3**, 107–110.
HUGUENIN, G. R., MANCHESTER, R. N., and TAYLOR, J. H. (1971). *Ap. J.* **169**, 97–104.
HULSE, R. A., and TAYLOR, J. H. (1974). *Ap. J.* **191**, L59–L61.
HUNT, G. C. (1971). *Mon. Not. Roy. Astron. Soc.* **153**, 199–131.
IEEE International Convention Digest (1974). Session 9, "CCDs in Analog Signal Processing" (in press).
JENKINS, G. M., and WATTS, D. G. (1968). "Spectrum Analysis and its Applications." Holden-Day, San Francisco, California.
KLAUDER, J. R., PRICE, A. C., DARLINGTON, S., and ALBERSHIEM, W. J. (1960). *Bell System Tech. J.* **39**, 745–808.
LANG, K. R. (1971). *Ap. J.* **164**, 249–264.
LANG, K. R., and RICKETT, B. J. (1970). *Nature (London)* **225**, 528–530.
LOVELACE, R. V. E., and CRAFT, H. D., JR. (1968). *Nature (London)* **220**, 875–879.
LOVELACE, R. V. E., SUTTON, J. M., and SALPETER, E. E. (1969). *Nature (London)* **222**, 231–233.
LYNE, A. G., and RICKETT, B. J. (1968). *Nature (London)* **218**, 326–330.
LYNE, A. G., SMITH, F. G., and GRAHAM, D. A. (1971). *Mon. Not. Roy. Astron. Soc.* **153**, 337–382.
MANCHESTER, R. N. (1972). *Ap. J.* **172**, 43–52.
MANCHESTER, R. N., and PETERS, W. L. (1972). *Ap. J.* **173**, 221–226.
MANCHESTER, R. N., and TAYLOR, J. H. (1972). *Astrophys. Lett.* **10**, 67–70.
MANCHESTER, R. N., and TAYLOR, J. H. (1974). *Ap. J.* **191**, L63–L64.
MANCHESTER, R. N., TADEMARU, E., TAYLOR, J. H., and HUGUENIN, G. R. (1973). *Ap. J.* **185**, 951–957.
MANCHESTER, R. N., TAYLOR, J. H., and VAN, Y. Y. (1974). *Ap. J.* **189**, L119–L122.
NELSON, J., HILLS, R., CUDABACK, D., and WAMPLER, J. (1970). *Ap. J.* **161**, L235–L244.
O'MEARA, T. R. (1960). *Proc. IEEE* **48**, 11, 1916–1918.
ORSTEIN, G. S. F. (1970). *Rev. Sci. Instrum.* **41**, 7, 957–959.
OSTRIKER, J. P., and GUNN, J. E. (1969). *Ap. J.* **157**, 1395–1417.
PAPALIOLIOS, C., CARLETON, N. P., and HOROWITZ, P. (1970). *Nature (London)* **228**, 445–450.
RADHAKRISHNAN, V., and MANCHESTER, R. N. (1969). *Nature (London)* **222**, 228–229
RANKIN, J. M. (1974). Paper presented at the Stanford Pulsar Symposium, 1974.
RANKIN, J. M., and COUNSELMAN, C. C., III. (1973). *Ap. J.* **181**, 875–889.
RANKIN, J. M., COMELLA, J. M., CRAFT, H. D., JR., RICHARDS, D. W., CAMPBELL, D. B., and COUNSELMAN, C. C., III. (1970). *Ap. J.* **162**, 707–725.
RANKIN, J. M., CAMPBELL, D. B., and BACKER, D. C. (1974). *Ap. J.* **188**, 609–613.
REICHLEY, P. E., and DOWNS, G. S. (1969). *Nature (London)* **222**, 229–230.
RICHARDS, D. W. (1972). Ph.D. Thesis, Cornell University, Ithaca, New York.
RICHARDS, D. W., PETENGILL, G. A., COUNSELMAN, C. C., III, and RANKIN, J. M. (1970). *Ap. J.* **160**, L1–L6.
RICKETT, B. J. (1969). Ph.D. Thesis, Manchester University, England.
RICKETT, B. J. (1970). *Mon. Not. Roy. Astron. Soc.* **150**, 67–91.

RICKETT, B. J. (1974). Paper presented at Stanford Pulsar Symposium, 1974.
RICKETT, B. J. (1975). *Ap. J.* **197**, 185.
RICKETT, B. J., and HANKINS, T. H. (1973). *Bull. Amer. Astron. Soc.* **5**, 18.
SALPETER, E. E. (1969). *Nature (London)* **221**, 31–33.
SANDE, G. (1968). "Arbitrary Radix One-Dimensional Fast Fourier Transform Subroutine." Univ. of Chicago Press, Chicago, Illinois.
SANGSTER, F. L. J., and TEER, K. (1969). *IEEE J. Solid-State Circuits* **4**, B1–B6.
SCARGLE, J. D., and PACINI, F. (1971). *Nature (London) Phys. Sci.* **232**, 144–149.
SHITOV, YU. P. (1972). *Sov. Astron,—AJ* **16**, 383–392.
SQUIRE, W. D., WHITEHOUSE, H. J., and ALSUP, J. M. (1969). *IEEE Trans. Microwave Theory Tech.* **17**, 1020–1040.
STAELIN, D. H. (1969). *Proc. IEEE* **57**, 4, 724–725.
STAELIN, D. H., and SUTTON, M. J. (1970). *Nature (London)* **226**, 69–70.
SUTTON, J. M. (1971). *Mon. Nat. Roy. Astron. Soc.* **155**, 51–64.
SUTTON, J. M., STAELIN, D. H., PRICE, R. M., and WEINER, R. (1970). *Ap. J.* **159**, L89–L93.
TANENBAUM, B. S., ZEISSIG, G. A., and DRAKE, F. D. (1968). *Science* **160**, 760–761.
TAYLOR, J. H. (1973). *Proc. IEEE* **61**, 1295–1298.
TAYLOR, J. H. (1974a). *Astron. & Astrophys., Suppl.* **15**, 367.
TAYLOR, J. H. (1974b). Paper presented at the Stanford Pulsar Symposium, 1974.
TAYLOR, J. H., and HUGUENIN, G. R. (1971). *Ap. J.* **167**, 273–291.
TAYLOR, J. H., HUGUENIN, G. R., and HIRSCH, R. M. (1971). *Astrophys. Lett.* **9**, 205–208.
TAYLOR, J. H., MANCHESTER, R. N., and HUGUENIN, G. R. (1975). *Ap. J.* **195**, 513–528.
VANDENBERG, N. R., CLARK, T. A., ERICKSON, W. C., RESCH, G. M., BRODERICK, J. J., PAYNE, R. R., and YOUMANS, A. B. (1973). *Ap. J.* **180**, L27–L29.
VAUGHAN, A. E., and LARGE, M. I. (1969). *Proc. ASA* **5**, 220–223.
WIELEBINSKI, R., VAUGHAN, A. E., and LARGE, M. I. (1969). *Nature (London)* **221**, 47.
WILLIAMSON, I. P. (1972). *Mon. Not. Roy. Astron. Soc.* **157**, 55–71.

Aperture Synthesis

W. N. Brouw*

NETHERLANDS FOUNDATION FOR RADIO ASTRONOMY† AND
LEIDEN OBSERVATORY
LEIDEN, THE NETHERLANDS

I. Introduction 131
II. Aperture Synthesis 132
   A. Aperture Synthesis Theory 133
   B. Principal Aperture Synthesis Formulas 136
III. Earth Rotation Aperture Synthesis 139
   A. Theory of Rotational Synthesis 139
   B. Delays 141
   C. Some Practical Limitations 143
   D. Sensitivity 154
   E. Single Beam vs. Aperture Synthesis Telescope 158
IV. Data Processing 160
   A. Telescope Steering and Data Collection 160
   B. Calibration 161
   C. Correction 162
   D. Transformation of Correlation Function into Brightness Distribution . 163
   E. Map Handling 169
V. Conclusion 173
   References 173

I. Introduction

Since the detection of radio noise from the Galaxy by Karl G. Jansky in 1933, radio astronomers have tried to increase the resolving power of their instruments. Because radio wavelengths are relatively long compared to optical wavelengths, the resolving power of radio telescopes is not as good as that of their optical counterparts. The resolution of large optical telescopes is limited by the effects of atmospheric disturbances to about 1 sec of arc. The resolution of single antenna radio telescopes is always limited by the diffraction pattern. A 25 meter dish has at 1.4 GHz a resolving power of $0.5°$, or 2000 times worse than an optical telescope. Even the largest fully steerable radiotelescope presently existing,

---

* Present address: Radiosterrenwacht, Dwingeloo, The Netherlands.
† Operated with the financial support of the Netherlands Organization for the Advancement of Pure Research (Z.W.O.).

the 100 meter telescope of the Max-Planck-Institut in Germany has, even at its highest operation frequency of about 15 GHz, a resolving power fifty times worse than an optical telescope.

From the early days of radio astronomy, radio astronomers have, therefore, looked for ways to increase their resolving power. By 1963 fan beams of the order of 1 minute of arc were obtained by the use of interferometer techniques (e.g., Swarup et al., 1963), and Earth rotation synthesis using fan beams had been shown to work (Christiansen and Warburton, 1955). Rotational aperture synthesis, using the rotation of the Earth and dishes movable on tracks, capable of synthesizing a large area of the sky with high resolution was introduced by Cambridge radio astronomers in 1962 (Ryle, 1962). In addition, many compound interferometers using "T", cross, or ring-shaped layouts of elementary antenna elements had produced pencil beams with a resolution depending on the size of the large structure, which was of the order of 1 km. In 1967 the availability of very stable frequency sources made the development of very long baseline interferometry (VLBI) possible, in which the resolving power is at the moment only limited by the diameter of the Earth. However, the very costly post-processing necessary for VLBI observations makes this a not very suitable method for continuous observing (e.g., Cohen, 1973).

In the last decade the method of rotational aperture synthesis has been proved to be a very reliable and powerful instrument for continuous routine observations, and many have been built. The instrument with the longest baseline now operational is the 5 km array in Cambridge, England (Ryle, 1972). The construction of a synthesis telescope with a maximum baseline of 36 km has been started by the National Radio Astronomy Observatory in the United States.

In this paper a brief outline of the theory of aperture synthesis is given in Section II. In Section III the special case of rotational aperture synthesis and its problems is described. Section IV describes the data processing necessary to convert the parameters observed with an aperture synthesis telescope into an intensity distribution of part of the sky. Some concluding remarks are contained in Section V.

II. Aperture Synthesis

The theory of aperture synthesis is well established, and there exist several descriptions in the literature. The reader is referred to these for a full understanding of the theory involved, and for an insight in the many different ways the principles of aperture synthesis can be described. Rather physical explanations can be found in Ryle and Hewish (1960) and Christiansen and Högbom (1969), for example, while more theoretical papers are Bracewell (1958), Swenson and Mathur (1968), and Rogers (1968).

In the following section a brief theoretical introduction is given in terms of classical electromagnetic theory, without going into all the details a thorough discussion should require. Full details of all the pitfalls to be expected can be found in many textbooks (e.g., Born and Wolf, 1964) or in the references above.

A. APERTURE SYNTHESIS THEORY

The instantaneous electric field $\mathbf{E}(\mathbf{r}, t)$ at a time t and a position $\mathbf{r}$ in space, can be related by Fourier theory to spatial frequency components $\mathbf{e}(\mathbf{k}, \nu)$:

$$\mathbf{E}(\mathbf{r}, t) = \iint_{-\infty}^{+\infty} \mathbf{e}(\mathbf{k}, \nu) \exp[2\pi i(\nu t - \mathbf{k} \cdot \mathbf{r})] \, d\mathbf{k} \, d\nu \tag{1}$$

$$\mathbf{e}(\mathbf{k}, \nu) = \iint_{-\infty}^{+\infty} \mathbf{E}(\mathbf{r}, t) \exp[-2\pi i(\nu t - \mathbf{k} \cdot \mathbf{r})] \, d\mathbf{r} \, dt \tag{2}$$

Since $\mathbf{E}$ must satisfy the wave equations, it can easily be shown that in a lossless, source-free medium $|\mathbf{k}| = \nu/c$. $\mathbf{e}(\mathbf{k}, \nu)$ can be interpreted as the complex amplitude of a plane wave with a frequency ν, traveling in a direction with direction cosines $(c/\nu)\mathbf{k}$, where c is the velocity of light.

Equation (1) states that any electromagnetic field can be thought of as the superposition of plane waves (see, e.g., Stratton, 1941, p. 362). The seemingly physical impossibility of negative frequencies can be circumvented if we consider that $\mathbf{E}$ must be real. In that case we know that $\mathbf{e}$ must be hermetian, i.e., $\mathbf{e}(\mathbf{k}, \nu) = \mathbf{e}^*(-\mathbf{k}, -\nu)$, where the asterisk denotes a complex conjugate. Furthermore, we will only consider real values for the components of the vector $\mathbf{k}$, i.e., $\mathbf{e}$ will only be defined on a sphere with radius ν/c.

The restriction $|\mathbf{k}| = \nu/c$ means that Eq. (1) can be reduced to a three-dimensional, rather than a four-dimensional, integral. However, the phase term will be

$$\exp[2\pi i(\nu t - \mathbf{k} \cdot \mathbf{r})]$$
$$= \exp[2\pi i(\nu t - k_x r_x - k_y r_y - (\nu^2/c^2 - k_x^2 - k_y^2)^{1/2} r_z)] \tag{3}$$

For more easy handling I will, however, keep to the four-dimensional form with the understanding that $\mathbf{e}(\mathbf{k}, \nu)$ is only defined in points $\mathbf{k}$ lying on a sphere with radius ν/c and is equal to zero elsewhere.

For the same reason we can also say that if $\mathbf{E}$ is defined in all points $\mathbf{r}$ on a plane, the complex amplitudes $\mathbf{e}$ can be determined. In the special case of a plane with $r_z = 0$, we can rewrite Eqs. (1)–(2) as

$$\mathbf{E}(r_x, r_y, 0, t) = \iiint_{-\infty}^{+\infty} \mathbf{e}(\mathbf{k}, \nu) \exp[2\pi i(\nu t - k_x r_x - k_y r_y)] \, dk_x \, dk_y \, d\nu \quad (4)$$

$$\mathbf{e}(\mathbf{k}, \nu) = \iiint_{-\infty}^{+\infty} \mathbf{E}(r_x, r_y, 0, t) \exp[-2\pi i(\nu t - k_x r_x - k_y r_y)] \, dr_x \, dr_y \, dt \quad (5)$$

The intensity of the energy flow of a plane wave with a complex amplitude $\mathbf{e}(\mathbf{k}, \nu)$ is given by the average Poynting vector

$$\mathbf{S}(\mathbf{k}, \nu) = S(\mathbf{k}, \nu) \mathbf{k_u} = \tfrac{1}{2}(\epsilon/\mu)^{1/2} \, \mathbf{e}(\mathbf{k}, \nu) \cdot \mathbf{e}^*(\mathbf{k}, \nu) \cdot \mathbf{k_u} \quad (6)$$

where $\mathbf{k_u}$ is the unit vector in the direction $\mathbf{k}$, and ϵ and μ are the dielectric constant and permeability of the medium. $S(\mathbf{k}, \nu)$ is related to space-time coordinates by

$$S(\mathbf{k}, \nu) = \tfrac{1}{2}(\epsilon/\mu)^{1/2} \iint_{-\infty}^{+\infty} \{\mathbf{E}(\mathbf{r}, t) * \mathbf{E}^*(-\mathbf{r}, -t)\}$$
$$\times \exp[-2\pi i(\nu t - \mathbf{k} \cdot \mathbf{r})] \, d\mathbf{r} \, dt \quad (7)$$

where the function asterisk denotes convolution, and the term in curly brackets describes the autocorrelation function of $\mathbf{E}(\mathbf{r}, t)$, defined as

$$\mathbf{E}(\mathbf{r}, t) * \mathbf{E}^*(-\mathbf{r}, -t) = \iint_{-\infty}^{+\infty} \mathbf{E}(\mathbf{r}', t') \cdot \mathbf{E}^*(\mathbf{r}' - \mathbf{r}, t' - t) \, d\mathbf{r}' \, dt' \quad (8)$$

Note that $\mathbf{E}$ is real, and $\mathbf{E}^*$ is only written to obtain symmetry in the formulas. The quantity of interest to radio astronomy is the brightness, i.e., the power per unit area, per unit solid angle, and per frequency interval. Defining $\mathbf{k} = (\nu/c)\mathbf{k_u} = (\nu/c)(l, m, n)$, where l, m, and n are direction cosines, and noting that the solid angle subtended by an element $dl \, dm$ is equal to $n^{-1} \, dl \, dm$, we find that we can define the brightness B as

$$B(\mathbf{k}, \nu) = \tfrac{1}{2} n^{-1} (\nu/c)^2 (\epsilon/\mu)^{1/2} \iint_{-\infty}^{+\infty} \{\mathbf{E}(\mathbf{r}, t) * \mathbf{E}^*(-\mathbf{r}, -t)\}$$
$$\times \exp[-2\pi i \nu/c(ct - \mathbf{k_u} \cdot \mathbf{r})] \, d\mathbf{r} \, dt \quad (9)$$

Hence, if we can determine the autocorrelation function of the instantaneous electric field, the brightness distribution can be determined. We can go even further if we bear the restrictions in mind that were mentioned before, and state that if we determine the autocorrelation function of the instantaneous electric field on any plane, we can determine the brightness distribution.

In practice, however, it is impossible to measure this autocorrelation function. However, if we assume ergodic circumstances, we can expect that

$$\int\int_{-\infty}^{+\infty} \mathbf{E}(\mathbf{r}', t') \cdot \mathbf{E}^*(\mathbf{r}' - \mathbf{r}, t' - t) \, d\mathbf{r}' \, dt' = \lim_{T \to \infty} \frac{1}{2T} \int_{-T}^{+T} \mathbf{E}(\mathbf{r}', t') \cdot \mathbf{E}^*(\mathbf{r}' - \mathbf{r}, t' - t) \, dt' \quad (10)$$

Physically this means that the time average of the product in Eq. (10) will be independent of the actual position and time at which it is measured, but will only depend on the relative position and time of the two points at which we sample the electric field.

Hence, instead of having to take the spatial average of the product of the electric field intensities at two points at a distance $(\mathbf{r}' - \mathbf{r}, t' - t)$, we only have to take the time average of any two points at a separation $(\mathbf{r}' - \mathbf{r}, t' - t)$. The assumption of ergodicity is a basic assumption of aperture synthesis as explained above. It corresponds to the case of completely incoherent radiation in other descriptions of the basic aperture synthesis theory.

If we write

$$P(\mathbf{r}, t) = \lim_{T \to \infty} \frac{1}{2T} \int_{-T}^{+T} \mathbf{E}(\mathbf{r}', t') \cdot \mathbf{E}^*(\mathbf{r}' - \mathbf{r}, t' - t) \, dt' \quad (11)$$

and use Eq. (9), we come to the basic formulas of aperture synthesis

$$B(\mathbf{k}, \nu) \propto n^{-1} \nu^2 \int\int_{-\infty}^{+\infty} P(\mathbf{r}, t) \exp[-2\pi i(\nu t - \mathbf{k} \cdot \mathbf{r})] \, d\mathbf{r} \, dt \quad (12)$$

or

$$B(\mathbf{k}_u, \nu) \propto n^{-1} \nu^2 \int\int_{-\infty}^{+\infty} P[(\nu/c)\mathbf{r}, t] \exp[-2\pi i[\nu t - (\nu/c)\mathbf{k}_u \cdot \mathbf{r}]] \, d\mathbf{r} \, dt \quad (13)$$

where we should explicitly note that $\mathbf{r}$ and t are only *differences* in coordinates between two sample points. This has the immediate consequence that the values of $P(\mathbf{r}, t)$ for different $\mathbf{r}$ and t do not have to be measured concurrently, but can be observed one at a time.

B. Principal Aperture Synthesis Formulas

If we consider the final result of Section II, A, where we related the autocorrelation function of the instantaneous electric field P to the sky brightness distribution B in Eqs. (12)–(13), we can note the following:

1. Since P and B are real, we know that they must be symmetric, i.e.,

$$B(-\mathbf{k}, -\nu) = B(\mathbf{k}, \nu) \quad \text{and} \quad P(-\mathbf{r}, -t) = P(\mathbf{r}, t) \quad (14)$$

2. In practice we will sample the electric field by measuring the instantaneous voltage produced at the terminal wires of an element having a certain sensitivity $\mathbf{r}(\mathbf{k}, \nu)$ in the spatial-frequency domain, i.e.,

$$V(\mathbf{r}, t) = \mathbf{E}(\mathbf{r}, t) * \mathbf{R}(\mathbf{r}, t) \quad (15)$$

where $\mathbf{R}$ is the Fourier transform of $\mathbf{r}$.

If we measure the autocorrelation function of V with the use of two elements having sensitivities $\mathbf{r}_1$ and $\mathbf{r}_2$, the brightness distribution obtained by taking the Fourier transform of the voltage autocorrelation will be equal to

$$B_{\text{obs}}(\mathbf{k}, \nu) = B_{\text{true}}(\mathbf{k}, \nu) \, \mathbf{r}_1(\mathbf{k}, \nu) \cdot \mathbf{r}_2{}^*(\mathbf{k}, \nu) \quad (16)$$

3. If we measure the autocorrelation P only for $t = 0$, and if we determine a brightness distribution by taking the three-dimensional Fourier transform of $P(\mathbf{r}, t)$, we will observe

$$B_{\text{obs}}(\mathbf{k}) = \int_{-\infty}^{+\infty} B_{\text{true}}(\mathbf{k}, \nu) \, \mathbf{r}_1(\mathbf{k}, \nu) \cdot \mathbf{r}_2{}^*(\mathbf{k}, \nu) \, d\nu \quad (17)$$

This has two definite drawbacks:

a. A radio source in a direction $\mathbf{k}_u$, will have different $\mathbf{k}$'s for different frequencies ν, since $|\mathbf{k}| = \nu/c$. Hence, an integration over all frequencies at a specific $\mathbf{k}$ will average the contributions from sources in many directions $\mathbf{k}_u$. This difficulty can be overcome by making $\mathbf{r}(\mathbf{k}, \nu)$ a narrowband filter. In that case the integral in Eq. (17) has a meaning over a very limited range in frequency only.

b. However, even if our instrument is sensitive for one frequency ν_0 only, we note that:

$$B_{\text{obs}}(\mathbf{k}, \nu_0) = B_{\text{true}}(\mathbf{k}, \nu_0) + B_{\text{true}}(\mathbf{k}, -\nu_0) \quad (18)$$

which, according to Eq. (14) can be written as

$$B_{\text{obs}}(\mathbf{k}, \nu_0) = B_{\text{true}}(\mathbf{k}, \nu_0) + B_{\text{true}}(-\mathbf{k}, \nu_0) \quad (19)$$

Hence, it is impossible to determine the brightness distribution uniquely.

This difficulty can be overcome if we measure a complex autocorrelation function. This complex correlation function is obtained by the following, nonlinear, process. The real part is equal to the correlation function P as defined above. The imaginary part $P_i(\mathbf{r}, t)$ is equal to the function obtained by taking the average of the instantaneous field in one place multiplied by the instantaneous field at a distance $(\mathbf{r}, 4\nu^{-1})$. That is, we apply a phase shift of $\pi/2$ to all the elementary plane waves composing the field.

The Fourier transform of this complex correlation function will give the desired uniquely determined brightness distribution. It can easily be seen why this is the case. We would like to determine

$$B_1(\mathbf{k}, \nu) = H(\nu) B(\mathbf{k}, \nu) \tag{20}$$

where $H(\nu)$ is the Heaviside unit step defined as

$$H(\nu) = \begin{cases} 1 & \nu > 0 \\ 0 & \nu < 0 \end{cases} \tag{21}$$

In other words, we want to measure a brightness distribution for positive frequencies only. To obtain this, we should take the convolution of $P(\mathbf{r}, t)$ with the Fourier transform of $H(\nu)$

$$FT[H(\nu)] = \tfrac{1}{2}\delta(t) - i/2\pi t \tag{22}$$

where $\delta(t)$ is the Dirac delta function. Since convolution is distributive over addition, it suffices to take the convolution of Eq. (22) with only one of the electric fields used in producing the correlation function P. Let us take $\mathbf{E}(\mathbf{r}, t)$ at a distance $(\mathbf{r}, t)$ from the other. Since $\mathbf{E}(\mathbf{r}, t)$ is the integral of plane waves, it is sufficient, due to the distributive property of convolution, to consider the convolution of one of the constituent plane waves with Eq. (22); we get

$$\begin{aligned} \{\mathbf{e}(\mathbf{k}, \nu) & \exp[2\pi i(\nu t - \mathbf{k}\cdot\mathbf{r})]\} * \{\tfrac{1}{2}\delta(t) - i/2\pi t\} \\ &= \tfrac{1}{2}\mathbf{e}(\mathbf{k}, \nu) \exp[2\pi i(\nu t - \mathbf{k}\cdot\mathbf{r})] - i\, \mathbf{e}(\mathbf{k}, \nu) \exp[2\pi i(\nu t - \mathbf{k}\cdot\mathbf{r})] \\ &\quad \times \int_{-\infty}^{+\infty} \exp[-2\pi i\nu\, t']/2\pi t'\, dt' \\ &= \tfrac{1}{2}\mathbf{e}(\mathbf{k}, \nu) \exp[2\pi i(\nu t - \mathbf{k}\cdot\mathbf{r})] \\ &\quad - i\, \mathbf{e}(\mathbf{k}, \nu) \exp[2\pi i(\nu t - \tfrac{1}{4} - \mathbf{k}\cdot\mathbf{r})] \end{aligned} \tag{23}$$

In the remainder of this article we will only consider complex correlation functions $P(\mathbf{r})$ measured at $t = 0$ only. The correlation radio-spectrograph is described elsewhere in this volume (see the article by John A. Ball, p. 177). Since $B(\mathbf{k})$ is real we know that $P(\mathbf{r})$ is Hermitian, i.e.,

$$P(-\mathbf{r}) = P^*(\mathbf{r}) \qquad (24)$$

hence, to determine the brightness distribution $B(\mathbf{k})$ it suffices to measure only half of the complex correlation function.

4. If we measure the complex correlation function $P(\mathbf{r})$ on a plane in space only, the brightness distribution obtained by taking the two-dimensional Fourier transform of $P(x, y, 0)$ will be

$$B_{\text{obs}}(\mathbf{k}) = \int_{-\infty}^{+\infty} B_{\text{true}}(\mathbf{k}) \, dk_z \qquad (25)$$

However, since B at specific values of k_x and k_y is only determined for two points k_z, i.e., $k_z = \pm (\nu^2/c^2 - k_x^2 - k_y^2)^{1/2}$, the use of receiving elements with high directivity which measure only radiation coming from a direction close to one of the two possible k_z values will solve this problem. Combining all the remarks specified above we obtain the final formulas for a continuum aperture synthesis system which samples the complex autocorrelation function $P(x, y, 0; 0)$:

$$\begin{aligned} B_{\text{obs}}(\mathbf{k_u}) &= B_{\text{obs}}(l, m; \nu_0) \\ &\propto n^{-1} \int_{-\infty}^{+\infty} P_{\text{obs}}(x, y, 0; 0) \exp\left[2\pi i\left(x\, l\,\frac{\nu_0}{c} + y m\,\frac{\nu_0}{c}\right)\right] dx\, dy \\ &\propto n^{-1} \int_{-\infty}^{+\infty} P(U, V) \exp[2\pi i(U\, l + V\, m)]\, dU\, dV \\ &= \int_{-\infty}^{+\infty} \int_0^{+\infty} B_{\text{true}}(l, m, n; \nu)\, g(l, m, n; \nu)]\, d\nu\, dn \end{aligned} \qquad (26)$$

where
$$g(l, m, n; \nu) = \mathbf{r}_1(l, m, n; \nu) \cdot \mathbf{r}_2(l, m, n; \nu) \qquad (27)$$

and, in the case where g is a highly directive function in both frequency and space, Eq. (26) reduces to

$$B_{\text{obs}}(l, m; \nu_0) = B_{\text{true}}(l, m, n; \nu_0)\, g(l, m, n; \nu_0) \qquad (28)$$

It will be clear that the polarization characteristics of the observed radiation will depend on the polarization sensitivity of the receiving elements. To include all polarization information in the observed brightness distribution is a very simple extension of the basic theory. The reader is referred to articles on this aspect for further reference (e.g., Ko, 1967; Weiler, 1973).

III. Earth Rotation Aperture Synthesis

Cambridge University radio astronomers (Ryle et al., 1965) have pioneered a technique especially suited for the synthesis of large effective apertures. In its simplest form such an instrument consists of two identical antennas. Both antennas track the source during an observation. Due to the rotation of the Earth the relation of the baseline of the interferometer changes with respect to the sky, hence, the complex correlation function of the instantaneous electric field will be measured at many different positions during an observing run. If one of the two antennas of the interferometer can be moved with respect to the other, the procedure can be repeated as often as necessary with a different baseline to obtain more samples of the correlation function. By careful selection of the relative positions of the two antennas, enough points of the correlation function can be sampled to construct a map of the area of sky under investigation. This technique has been named "super-synthesis" by the Cambridge group. However, the term Earth rotation or rotational aperture synthesis seems more appropriate.

A. Theory of Rotational Synthesis

The customary astronomical coordinate system is defined by the instantaneous rotational axis of the Earth. Right ascensions (α) are measured in the equatorial plane in a counter clockwise direction from one of the intersections of this plane with the ecliptic plane, i.e., the plane defined by the Earth's orbit. Declinations (δ) are measured from the equator plane toward the pole of the rotational axis. In this system the direction toward a celestial source ($\mathbf{k_u}$) is defined as

$$\mathbf{k_u} = \begin{pmatrix} l \\ m \\ n \end{pmatrix} = \begin{pmatrix} \cos \alpha \; \cos \delta \\ \sin \alpha \; \cos \delta \\ \sin \delta \end{pmatrix} \tag{29}$$

while the interferometer is specified by

$$\mathbf{R} = \begin{pmatrix} U \\ V \\ W \end{pmatrix} = R \begin{pmatrix} \cos(\theta - t_p) \; \cos \delta_p \\ \sin(\theta - t_p) \; \cos \delta_p \\ \sin \delta_p \end{pmatrix} \tag{30}$$

where R = distance between interferometer elements in wavelengths, θ = local apparent sidereal time, δ_p = declination of line joining interferometer elements, $t_p = \theta - \alpha_p$ = hour-angle of line joining interferometer elements, and α_p = right ascension of line joining interferometer elements.

R, t_p, and δ_p are fixed for a given interferometer, and hence during a (sidereal) day α_p will vary continuously, thus causing a continuous change of the baseline **R** with respect to the celestial sphere. The endpoint of the vector **R** will trace out circles in a plane parallel to the (U, V) plane at a distance $R \sin \delta_p$ from it. The radius of the circle is $R \cos \delta_p$. The output of a complex correlatorreceiver for a monochromatic point source with an intensity B ($\mathbf{k_u}$, ν_0) will be (cf. Section II)

$$P(\mathbf{R}) = B(\mathbf{k_u}, \nu_0) \exp[-2\pi i \mathbf{k_u} \mathbf{R}] \tag{31}$$

In the case of rotational synthesis the phase of P will vary with time, giving rise to so-called "fringes." These fringes will have frequencies

$$\nu_f = \frac{d \arg(P(\mathbf{R}))}{dt} = -2\pi \mathbf{k_u} \, (d\mathbf{R}/dt) \tag{32}$$

where

$$\frac{d\mathbf{R}}{dt} = R \begin{pmatrix} -\sin(\theta - t_p) \cos \delta_p \\ \cos(\theta - t_p) \cos \delta_p \\ 0 \end{pmatrix} \frac{d\theta}{dt} \tag{33}$$

The fringes have to be recorded several times per fringe. However, in general the interferometer elements are sensitive only for radiation coming from a small area in the sky. Hence, the range in fringe frequencies, at any one time, will be small. Most synthesis telescopes, therefore, have one way or another of so-called "fringe stopping." This means that the output of the complex correlator is mixed with a frequency corresponding to the fringe frequency of a specific position on the sky, the so-called fringe-stopping center. This fringe-stopping center is, in general, the position on the sky pointed at by the optical axis of the interferometer elements. In that case we have a coordinate system in which

$$\mathbf{k}_u = \begin{pmatrix} l \\ m \\ n \end{pmatrix} - \begin{pmatrix} l_0 \\ m_0 \\ n_0 \end{pmatrix} = \begin{pmatrix} \cos \alpha \cos \delta - \cos \alpha_0 \cos \delta_0 \\ \sin \alpha \cos \delta - \sin \alpha_0 \cos \delta_0 \\ \sin \delta - \sin \delta_0 \end{pmatrix} \tag{34}$$

and

$$\mathbf{R} = R \begin{pmatrix} \cos(\theta - t_p) \cos \delta_p \\ \sin(\theta - t_p) \cos \delta_p \\ \sin \delta_p \end{pmatrix} \tag{35}$$

where the subscript 0 indicates the fringe-stopping center coordinates. The differential frequencies thus obtained can be recorded less frequently, which reduces the amount of data to be handled by a factor of about $\lambda \cos \delta_0 / D$. Here λ is the observing wavelength and D is the diameter of the telescope element. Physically it means that the origin of the sky coordinate system is displaced from the center of the reference celestial sphere to a point on the celestial sphere.

It is often advisable to have a sky coordinate system in such a way that the l and m coordinates correspond roughly to the standard celestial coordinate differences $\Delta\alpha \cos \delta$ and $\Delta\delta$. Such a system is obtained by rotating the system defined by Eq. (34) through an angle α_0 around the n axis, and dividing the m coordinate by $\sin \delta_0$. The system obtained has

$$\mathbf{k} = \begin{pmatrix} l \\ m \\ n \end{pmatrix} = \begin{pmatrix} -\sin(\alpha - \alpha_0) \cos \delta \\ [\cos(\alpha - \alpha_0) \cos \delta - \cos \delta_0]/\sin \delta_0 \\ -\sin \delta + \sin \delta_0 \end{pmatrix}$$

$$\mathbf{R} = \begin{pmatrix} U \\ V \\ W \end{pmatrix} = R \begin{pmatrix} \sin(t_p - t_0) \cos \delta_p \\ \cos(t_p - t_0) \cos \delta_p \sin \delta_0 \\ -\sin \delta_p \end{pmatrix} \quad (36)$$

A rotational aperture synthesis telescope consists generally of several elements which are connected to produce several values of the complex autocorrelation at the same time, or, has movable elements to allow for the measurement of the correlation function at many points R at different times, or is a combination of both.

If all the elements are on a line, or if the elements can be moved along a straight line only, the correlation function P will be sampled at points on a cone defined by Eq. (36) for variable R. Only in the case where all telescopes are on a straight line in an East–West direction (i.e., $\delta_p = 0$), or if the elements are arranged in such a way that the product $R \sin \delta_p$ remains constant, will the cone degenerate into a plane. Only in that case will a two-dimensional Fourier transform suffice to obtain the brightness distribution in the sky (cf. Section II, B). In all other cases a three-dimensional Fourier transform is necessary, or a two-dimensional transform that is not a proper Fourier transform (cf. Bracewell, 1965).

B. Delays

The interferometer systems used to sample the complex correlation function will, in practice, not be monochromatic, but will have a finite bandwidth. Indeed, they will have as wide a bandwidth as possible to obtain a high sensitivity. However, if we sample the correlation function only for one time difference, it has been shown in Section II, B that the brightness distribution obtained by taking the Fourier transform of the correlation function sampled in this way is equal to the integral of the brightness distribution over the finite bandwidth of the receiving elements. That is,

$$B(l, m, n, \nu_0) = \int_{\nu_0-\Delta\nu/2}^{\nu_0+\Delta\nu/2} B(l, m, n, \nu) \, d\nu \qquad (37)$$

This seems to be a quite legitimate procedure for continuum radiation as long as the bandwidth $\Delta\nu$ is not too large. However, one should bear in mind that the fringe stopping was done for one, central, frequency ν_0. For other frequencies the fringe-stopping center will be displaced by $\nu/\nu_0\, \mathbf{k}_{u,0}$, if $\mathbf{k}_{u,0}$ are the coordinates of the fringe-stopping center in the original coordinate system. Even for a very narrow bandwidth this will give an appreciable drift of the fringe-stopping center in the band, and hence, Eq. (37) will produce the integral over an area of the sky many synthesized beamwidths large. This problem can be solved if the fringe stopping is done the right way, i.e., with a frequency-dependent mixer frequency, or, in other words, with a time delay

$$\tau = \mathbf{k}_{u,0}\,(R/\nu_0). \qquad (38)$$

In practice it is impossible to insert a continuously varying time delay with a high enough precision at the signal frequency in the receiving equipment. Therefore, the time delay is inserted in the intermediate frequency part of the receiver, usually in steps corresponding to a full wavelength of the intermediate frequency. This produces, for the center of the band, no phase changes, and hence no fringe stopping. The fringe stopping is then done separately.

The net effect of this double procedure is to produce a fringe stopping with the right properties. The noncontinuous change in the delay produces a frequency-dependent phase error for an error setting $\Delta\tau$ of the delay

$$\Delta\phi = \Delta\tau(\nu - \nu_0) \qquad (39)$$

Hence, as long as $\Delta\tau\,\Delta\nu \ll 1$ the system will operate perfectly. Equation (39) states that for a given bandwidth $\Delta\nu$ a receiver consisting of separate delay and fringe-stopping mechanisms will operate all right if the steps in the setting of the delay are chosen as $\Delta\tau \ll (\Delta\nu)^{-1}$. However, even in the case of perfect fringe stopping, the choice of the bandwidth is not a free parameter, since the sky coordinates with respect to the fringe-stopping center themselves are a function of the frequency. To obtain a useful result from the integration in Eq. (37) the distance to the fringe-stopping center ρ, the bandwidth $\Delta\nu$, and the baseline R should be related by

$$\rho(\Delta\nu) \ll c/R \qquad (40)$$

if a map of the sky up to a distance ρ is wanted.

C. Some Practical Limitations

1. Limited Sampling

A practical rotational synthesis telescope will sample the autocorrelation function up to a maximum baseline only. Furthermore, the sample points will not cover the cone surface(s) up to this maximum baseline contiguously, but rather be limited to discrete points.

To get an idea of the influence of this discrete sampling on the observed brightness distribution, we will define

$$G(\mathbf{R}) = \sum_i \delta(\mathbf{R} - \mathbf{R}_i) \tag{41}$$

where the $\mathbf{R}_i$'s are all the sampled points. Writing $g(\mathbf{k})$ for the Fourier transform of $G(\mathbf{R})$ we have

$$P_{\text{obs}}(\mathbf{R}) = G(\mathbf{R}) \, P(\mathbf{R}) \tag{42}$$

and hence

$$B_{\text{obs}}(\mathbf{k}) = g(\mathbf{k}) * B(\mathbf{k}) \tag{43}$$

Hence, we can consider $g(\mathbf{k})$ as a synthesized beam pattern. In practice, for synthesis telescopes consisting of elements on a line which have sampled the correlation function along a set of regularly spaced ellipses, the beam will be equal to a central beam with a width depending on the maximum baseline sampled and a set of regularly spaced elliptical rings. The spacing between the rings is inversely proportional to the spacing between the sampling ellipses. The amplitude of the rings goes roughly as the square root of the ring number with respect to the center (Bracewell and Thompson, 1973). It is often convenient to include a "taper" in the sampling function, i.e.,

$$G(\mathbf{R}) = \sum_i T(\mathbf{R}) \, \delta(\mathbf{R} - \mathbf{R}_i) \tag{44}$$

The tapering function may be complex, of course, to include phase weighting. The taper function can, for instance, be used to lower the sidelobe level of the beam by giving lower weight to the longer baselines, although, of course, the width of the mainbeam and of the "grating rings" will be increased at the same time. It depends on the kind of observation done (e.g., point source or extended region) whether resolving power or low sidelobe level is the more advantageous.

By including all known errors in the weight function $G(\mathbf{R})$ the synthesis antenna pattern can always be calculated, and the interpretation of the brightness map obtained will be much easier.

It should be noted that only half of the conal surface has to be observed. Since the sky brightness is real, the complex correlation function has the property $P(-\mathbf{k}) = P^*(\mathbf{k})$. Hence a twelve hours observation suffices to describe an ellipse on the cone completely (cf. Section II, B).

2. Aspects of the Sky

The rotational axis of the Earth is not fixed in space, but is, due to the influence of other bodies in the planetary system, continuously changing its direction. The amount of change is given by the astronomical precession and nutation. Furthermore, the velocity of the interferometer elements with respect to the sources under observation is continuously changing due to the movement of the Earth around the center of gravity of the solar system (annual aberration), rotation of the Earth (diurnal aberration), and the relative motion of the Sun and the source (planetary aberration). The last term can be considered to be constant, except for bodies in the planetary system. The surface of the Earth to which the interferometer elements are fixed changes position with respect to the axis of rotation of the Earth. Part of this is due to the finite elasticity of the Earth (polar motion), while, furthermore, the influences of the Sun and Moon produce tidal effects (Earth tides). Erratic changes in the rotation rate of the Earth at the level of about 10^{-7} occur as well.

a. Aberration. If the velocity of the two interferometer elements is the same, and for all but very long baseline interferometry they can be considered to be so, the aberration causes a displacement $\Delta\mathbf{k}$ with respect to an inertial frame of reference. Correction of this shift can be fully made by incorporating an effect $\Delta\mathbf{k}\cdot\mathbf{r}$ in the fringe-stopping procedure. If the aberration $\Delta\mathbf{k}$ is known, insertion of this term in the fringe stopping (and delay) will produce results as if the observation were done in the inertial frame of reference.

b. Precession and Nutation. Precession and nutation change the initial frame of reference to another instantaneous frame of reference. The effect can be described by a rotation matrix $||S||$:

$$\mathbf{k}_{\text{inst}} = ||S||\ \mathbf{k}_{\text{init}} \qquad (45)$$

In the case where a two-dimensional Fourier transform suffices, the effect can never be described by a linear transformation. Hence, the Fourier transform will be converted to either a three-dimensional Fourier transform, or to a two-dimensional non-Fourier transform. During a 12-hour observation the effect can, in all practical cases, be considered as a shift $\Delta\mathbf{k}$ for the whole field under consideration, and the effects can be handled in the same way as for aberration. For observations which are separated by longer periods a first-order approximation can be made in the following way. Let us consider an (l, m) plane on which the brightness distribution is defined by a two-dimensional Fourier transform of the correlation values measured on a plane in the **r** space.

$$\boldsymbol{\tau}_e = \begin{pmatrix} l \\ m \end{pmatrix}_{\text{init}} \tag{46}$$

$$\boldsymbol{\tau}_d = \begin{pmatrix} l \\ m \end{pmatrix}_{\text{inst}} \tag{47}$$

$$\|R\| = \begin{pmatrix} S_{11}^{-1} - S_{13}^{-1} \cos \alpha_{0,d} \cot \delta_{0,d} & S_{12}^{-1} - S_{13}^{-1} \sin \alpha_{0,d} \cot \delta_{0,d} \\ S_{21}^{-1} - S_{23}^{-1} \cos \alpha_{0,d} \cot \delta_{0,d} & S_{22}^{-1} - S_{23}^{-1} \sin \alpha_{0,d} \cot \delta_{0,d} \end{pmatrix} \tag{48}$$

$$\|P\| = \begin{pmatrix} \sin \alpha_{0,e} & -\cos \alpha_{0,e} \\ \cos \alpha_{0,e}/\sin \delta_{0,e} & \sin \alpha_{0,e}/\sin \delta_{0,e} \end{pmatrix} \tag{49}$$

where the subscript 0 indicates values for the fringe-stopping center, and l and m are coordinates as defined by Eq. (36).
Then

$$\boldsymbol{\tau}_e = \|P\| \, \|R\| \, \boldsymbol{\tau}_d' \tag{50}$$

or

$$\begin{pmatrix} U_e \\ V_e \end{pmatrix} = \|P^{-1}\|^T \, \|R^{-1}\|^T \begin{pmatrix} U_d' \\ V_d' \end{pmatrix} \tag{51}$$

where the primed coordinates are those as defined by Eq. (35).

The error made by converting the instantaneous coordinates (subscript d) into the initial coordinates (subscript e) is equal to

$$\Delta \boldsymbol{\tau} = \begin{pmatrix} \left[-\frac{(\delta - \delta_0)^2}{2} - \frac{(\alpha - \alpha_0)^2}{2} \cos^2 \delta_0 + \text{higher order terms} \right] \\ \times (S_{13}^{-1} \sin \alpha_0 - S_{23}^{-1} \cos \alpha_0) \\ \left[-\frac{(\delta - \delta_0)^2}{2} - \frac{(\alpha - \alpha_0)^2}{2} \cos^2 \delta_0 + \text{higher order terms} \right] \\ \times (S_{13}^{-1} \cos \alpha_0 + S_{23}^{-1} \sin \alpha_0)/\sin \delta_0 \end{pmatrix} \tag{52}$$

c. Polar Motion. The interferometer baseline **R** is the difference between the positions **x** of the two individual elements. Polar motion is measured in the components x_p and y_p of the intersection of the instantaneous rotational axis with a reference plane perpendicular to an assumed initial rotational axis in angular measure. If the telescope coordinates are given by $\mathbf{x}_e$ in the initial frame of reference,

$$\mathbf{x}_e = R \begin{pmatrix} \cos t_e \cos \delta_e \\ -\sin t_e \cos \delta_e \\ \sin \delta_e \end{pmatrix} \tag{53}$$

where δ and t are the declination and hour-angle of the line from a coordinate origin to the interferometer element at a distance R from the origin, then the instantaneous position $\mathbf{x}_d$ is given by

$$\mathbf{x}_d = \begin{pmatrix} 1 & 0 & -x_p \cos \lambda_0 + y_p \sin \lambda_0 \\ 0 & 1 & x_p \sin \lambda_0 + y_p \cos \lambda_0 \\ x_p \cos \lambda_0 - y_p \sin \lambda_0 & -x_p \sin \lambda_0 - y_p \cos \lambda_0 & 1 \end{pmatrix} \mathbf{x}_e \tag{54}$$

where λ_0 is the initial East longitude of the coordinate origin. The $\mathbf{R}$ coordinate as defined by Eq. (36) between two interferometer elements 1 and 2 is then equal to

$$\mathbf{R} = \begin{pmatrix} -\sin t_0 & -\cos t_0 & 0 \\ \cos t_0 & -\sin t_0 & 0 \\ 0 & 0 & -1 \end{pmatrix} (\mathbf{x}_{1,d} - \mathbf{x}_{2,d}) \tag{55}$$

where t_0 is the hour-angle of the fringe-stopping center.

d. *Earth Tides.* The effect of Earth tides is, in general, less than a few times 10^{-7} of the baseline involved, and, in all cases where the baseline does not exceed 10 km, of no importance.

3. Atmosphere

Observing with an interferometer has a very important advantage over single-dish observations. The influence of the atmosphere and ionosphere on observations with a single telescope depends on the total atmosphere and ionosphere the radiation traverses. On the other hand, the effects on an interferometer observation depends on the difference between the atmospheres and ionospheres through which the radiation to the two elements passes. Of course, for sensitivity reasons, the pointing of the individual interferometer elements should be corrected for the total refraction. However, the precision with which the refraction ought to be known is of the order of the resolution of the individual elements, rather than the resolution of the interferometer. Refraction causes a phase-lag difference because of the curvature of the Earth's atmosphere. Even for a perfectly homogeneous atmosphere, the ray path from the two telescopes will pass through a differently scaled atmosphere. Corrections for this effect can be implemented by considering the effect for a simple model atmosphere. A model which considers a constant density atmosphere up to the reduced height and a refractivity depending on temperature, humidity, and pressure can attain a precision of about 0.5 mm at elevations of about 5°. Even for observations at about 1° from the horizon a precision of about 2 mm can be obtained (e.g., Altenhoff, 1966). Another cause for differential phase lag is atmospheric irregularity. Different ray paths for the different element telescopes will pass through different atmospheric irregularities. Tests have shown (Hinder and Ryle, 1971) that these irregularities have a correlation length of about 1 km. This puts a weather-dependent upper limit on the precision in phase that can be obtained. Hinder and Ryle (1971) find that at a baseline of 5 km a maximum resolution of about 0.1–1 arcsec can be obtained. The major influence of atmospheric irregularities will, however, be in the sidelobes of the synthesized beam pattern.

At very long baselines, where the atmospheres above the two elements are completely uncorrelated, only numerical ray tracing through model atmospheres can give a reliable estimate of the phase lag. It looks as if an absolute precision of 10 to 20 cm is at the moment attainable (Hemenway, 1974).

At frequencies below about 1 GHz ionospheric refraction plays a role in aperture synthesis as well. However, for the normal aperture synthesis where baselines never exceed a few tens of kilometers, a simple model of the ionosphere based on observed total electron contents will adequately correct for the effects down to about 100 MHz. In addition to the effects on the phase of observations, atmospheric extinction plays a role as well. As long as the frequency is below the region where absorption lines of water start to play a role (say, up to about 10 GHz) the simple curved atmosphere model described above will be able to correct for extinction to better than 1% of the signal at an elevation of 5° above the horizon.

4. Instrumental Imperfections

To be able to give a few quantitative results for some of the imperfections occurring in rotational aperture synthesis telescopes, in the remaining part of this section only East–West interferometers are considered. During a 12 hour observing period the complex correlation function $P(U, V)$ will be sampled on half an ellipse in the (U, V) plane. For computational ease a coordinate system (u, v) will be considered, in which the ellipse is transformed to a circle. We then have

$$\begin{aligned} u &= U = R \cos t_0 \\ v &= V/\sin \delta_0 = R \sin t_0 \\ L &= l = -\sin (\alpha - \alpha_0) \cos \delta \\ M &= m \sin \delta_0 = \cos (\alpha - \alpha_0) \cos \delta - \cos \delta_0 \end{aligned} \qquad (56)$$

where R = baseline of the interferometer in wavelengths, α = right ascension of point in the sky, t = hour-angle of point in the sky, δ = declination of point in the sky, and the suffix 0 indicates values for the fringe-stopping center (cf. Section III, A).

It will sometimes be convenient to use radial coordinates. In that case

$$R = (u^2 + v^2)^{1/2} \quad t = \text{arc tan } (V/U)$$
$$\rho = (L^2 + M^2)^{1/2} \quad \phi = \text{arc tan } (M/L) \quad (57)$$

will be used.

The Fourier transform of an error-free observation will produce a synthesized beam pattern $g(L, M)$ or $g(\rho, \phi)$. Since Fourier transformation is a linear operation instrumental errors can be considered to give a pattern $g + g_1$ or $g\, g_2$. In the case of a complete 12-hour observation with one interferometer we have

$$g(\rho, \phi) = J_0(2\pi R \rho) \quad (58)$$

where J_0 is the zero-order Bessel function of the first kind. A completely filled (u, v) plane up to a baseline $R_{\max}$ will give

$$g(\rho, \phi) = \frac{J_1(2\pi R_{\max} \rho)}{\pi R_{\max} \rho} \quad (59)$$

This pattern has a first sidelobe of about -13% and a half-power beamwidth of about $0.71/R_{\max}$ radians. To reduce this sidelobe level, but increase the width of the main peak, we can give a lower weight to the longer baselines. For example, if we use a weight function $w(R, t)$ defined by

$$w(R, t) = \text{arc cos } (r) - r\,(1 - r^2)^{1/2} \quad (60)$$

where $r = R/R_{\max}$ we get

$$g(\rho, \phi) = \frac{4[J_1(R_{\max} \rho)]^2}{\pi^2 R_{\max}^2 \rho^2} \quad (61)$$

with a half-power beamwidth of about $1.02/R_{\max}$ radians and a first sidelobe of about 1.2%.

An observation with many regularly spaced baselines (spacing ΔR) between $R_{\min}$ and $R_{\max}$ and a uniform weighting will have a beam pattern

$$g(\rho, \phi) = 1/N \sum_{n=0}^{N} (n + R_{\min}/\Delta R)\, J_0[2\pi\rho\,(n + R_{\min}/\Delta R)] \quad (62)$$

For an increasing number of baselines the main beam will approach

$$g_{\text{main}}(\rho, \phi) \approx \frac{R_{\max}\,J_1[2\pi(R_{\max} + \Delta R/2)\,\rho] - R_{\min}\,J_1[2\pi(R_{\min} - \Delta R/2)\,\rho]}{\pi\rho\,(R_{\max} + R_{\min})(R_{\max} - R_{\min} + \Delta R)} \quad (63)$$

Furthermore, a set of "grating rings" will appear where the sidelobes of the different J_0 functions defined by Eq. (62) will interfere constructively. These rings will be equally spaced in ρ at distances $1/\Delta R$. Bracewell and Thompson (1973) have shown that the grating ring at a distance $k/\Delta R$ can be very well represented near its maximum, in the case where $R_{\min} = 0$, by

$$g_k(\rho, \phi) \approx \frac{2}{\pi} (\pi \rho\, R_{\max})^{-1/2} \operatorname{sinc}^{(1/2)} [2\rho\, (R_{\max} + \Delta R/2)$$
$$- 2k\, (R_{\max} + \Delta R/2)/\Delta R] \quad (64)$$

where $\operatorname{sinc}^{(1/2)}(x)$ denotes the half-order derivative of the function $\sin(\pi x)/\pi x$.

Equation (63) shows a very important fact for aperture synthesis telescopes. Interferometers have a minimum baseline due to the fact that the individual elements cannot overlap in space. This means that the central area of the (u, v) plane cannot be observed. Equation (63) shows that the resulting main beam antenna pattern has, in addition to the expected central peak, a very broad, low intensity negative level. This is due to the fact that if we assume values of zero for nonobserved values of the correlation function, i.e., if we put zero for the unobservable $P(0, 0)$, the integral over the synthesized antenna pattern must be zero. A direct consequence of this is that regions in the sky with an area comparable to the primary beam pattern of the element telescopes will not show up on the synthesis map. The brightness scale of a map is, therefore, dependent on the extent of the region under consideration. This problem can only be solved correctly if the missing information is added, e.g., by observing the area of sky with a single telescope with a diameter equal to or larger than the diameter of the missing (u, v) plane area.

a. Phase Errors. The interferometer elements and the receiving equipment will never be ideal, hence an instrumental phase error will always be present. This "collimation error" can be measured, but a small residual error α can easily exist. If $\alpha \ll 1$, we can say

$$g_1(L, M) \approx -g(L, M) * [\alpha/\pi L] \quad (65)$$

if the 12-hour observation covered a u-half of the (u, v) plane. This will produce sidelobes with the same width as the main beam in the M direction, but produce a pair of antisymmetric sidelobes on both sides of the main beam in the L direction.

In the case of a single interferometer observation, we can write that the total pattern $G(L, M)$ is equal to

$$G(L, 0) = \cos \alpha\, g(L, 0) - \sin \alpha\, H_0(2\pi \rho\, R)$$

and
$$G(0, M) = \cos \alpha \, g(0, M) \tag{66}$$

where $H_0(x)$ is the zero-order Struve function (Abramowitz and Stegun, 1965).

If we have a baseline-dependent phase error $\alpha = \alpha_0 \, (R/R_{\max})$, which can occur in practice if the declination of the interferometer baseline was not determined correctly, we have

$$g_1(\rho, \phi) \approx -\frac{\alpha_0^2}{2} \frac{J_2(2\pi\rho \, R_{\max})}{(2\pi\rho \, R_{\max})^2}$$
$$- 2\alpha_0 \, R_{\max}^{-3} \int_0^{R_{\max}} R^2 \int_{-\pi/2}^{\pi/2} \sin\{2\pi\rho \, R \cos(t - \phi)\} \, dt \, dR \tag{67}$$

For the axes we have, in the case of an observed u half-plane:

$$g_1(L, 0) \approx -\frac{\alpha_0^2}{2} \frac{J_2(2\pi L \, R_{\max})}{(2\pi L \, R_{\max})^2} - 2\pi\alpha_0 \, R_{\max}^{-3} \int_0^{R_{\max}} R^2 \, H_0(2\pi L \, R) \, dR$$

$$g_1(0, M) \approx -\frac{\alpha_0^2}{2} \frac{J_2(2\pi M \, R_{\max})}{(2\pi M \, R_{\max})^2} \tag{68}$$

Away from the axes g is not definable as a function of well-studied functions.

b. Discrete Delay Stepping. In an aperture synthesis telescope time delays are a necessary component if the bandwidth of the receiving system $\Delta\nu$ is not much smaller than the inverse of the light time along the baseline (see Section III, A). Normally, this time delay is inserted in discrete steps $\Delta\tau$ at the intermediate frequency stage of the receiver. It is convenient to make $\Delta\tau$ equal to $(\nu_m - \nu_0)^{-1}$, or in other words, make $\Delta\tau$ equal to one-wavelength steps at the center of the intermediate frequency band. (ν_m is the L.O. frequency and ν_0 the central frequency of the band for which the receiving system is sensitive.) However, it is completely irrelevant, since we can always correct for the phase error at the central frequency, i.e., $2\pi \tau (\nu_m - \nu_0)$ in the fringe-stopping processor. It is convenient, furthermore, to insert delay lines in the paths from both telescopes, in such a way that the difference in the delay lengths equals the delay length required. For multiple element telescopes, this is more economic. Let us assume we put delays τ_1 and τ_2 in the two elements in such a way that each differs never by more than $\Delta\tau/2$ from the ideal value that can be calculated for each element. In the delay system a phase error $2\pi (\nu_m - \nu)(\tau_1 - \tau_2)$ will be produced. If we further correct for the error at the center frequency by correcting the phase of the output signal with $2\pi(\nu_m - \nu_0)(\tau_1 - \tau_2)$, the resulting phase error as a function of the frequency will be

$$\Delta\phi_1 = 2\pi(\nu - \nu_0)(\tau_1 - \tau_2) \tag{69}$$

The fringe-stopping mechanism introduces a phase rotation $2\pi\nu_0\tau$, where τ is the exact delay wanted. It should be noted that in principle a continuous delay switching at the signal frequency would solve the problem. However, in most practical systems the splitting into delay and fringe stopping is made. $2\pi\nu\,\tau$ should be inserted, and hence an error is made:

$$\Delta\phi_2 = 2\pi(\nu - \nu_0)\tau \tag{70}$$

The total phase error, as a function of frequency, due to the discrete delay stepping and the splitting into delay and fringe stopping of the wanted correction, is equal to

$$\begin{aligned}\Delta\phi &= 2\pi(\nu - \nu_0)(\tau - \tau_1 + \tau_2) \\ &= 2\pi\nu'\,(\tau - \tau_1 + \tau_2)\end{aligned} \tag{71}$$

In the case of an East–West interferometer, the delay for element i is calculated by

$$\tau_i = T_i \cos\delta_0 \sin t_0 + \text{constant} \tag{72}$$

where T_i is the light time of a telescope element with respect to a convenient zero point on the line connecting the interferometer elements, and the constant is chosen in such a way that τ_i is always positive. We can rewrite Eq. (72) as

$$\tau_i = T_i\,R^{-1}\,\nu_0^{-1} \cos\delta_0\,v + \text{constant} \tag{73}$$

where R is the baseline of the interferometer in wavelengths at frequency ν_0. The error made by the discrete delay stepping is equal to a sawtooth function which is zero at the points

$$v = k\,\Delta\tau\,R\,T_i^{-1}\,\nu_0 \sec\delta_0 \quad (k = 0, \pm1, \ldots) \tag{74}$$

and will vary between $-\Delta\tau/2$ and $+\Delta\tau/2$. The sawtooth function can be expressed as

$$\Delta\tau_i = [\alpha_i\,r\,v\,\Pi(\alpha_i\,v)] * \mathrm{III}(\alpha_i\,v)\,\alpha_i \tag{75}$$

where

$$\begin{aligned}\alpha_i &= \Delta\tau\,R\,T_i^{-1}\,\nu_0 \sec\delta_0 \\ \Pi(x) &= 0 \quad |x| > \tfrac{1}{2} \\ &= 1 \quad |x| < \tfrac{1}{2}\end{aligned}$$

$$\mathrm{III}(x) = \sum_{k=-\infty}^{+\infty} \delta(x-k)$$

Hence, the phase error can be written as

$$\Delta\phi = 2\pi\nu' (\Delta\tau_1 - \Delta\tau_2) \qquad (76)$$

The observed correlation function will be

$$P_{obs}(u, v) = P(u, v) \exp[i \Delta\phi] \qquad (77)$$

In the practical case, where the delay step $\Delta\tau \ll (\Delta\nu)^{-1}$ (the inverse of the bandwidth), and $\Delta\phi \ll 1$

$$P_{obs}(u, v) \simeq P(u, v)[1 + i \Delta\phi]$$
$$= P(u, v)[1 + 2\pi i \nu'(\Delta\tau_1 - \Delta\tau_2)] \qquad (78)$$

Hence, the error pattern will be equal to

$$g_1(L, M) = g(L, M) * [g_{11}(L, M) + g_{12}(L, M)] \qquad (79)$$

where $g_{1i}(L, M)$ is the Fourier transform of $2\pi i \nu' \Delta\tau_i$, which is equal to

$$g_{1i}(L, M) = -\delta(L) \frac{\nu' \Delta\tau}{M} \left[\cos\left(\frac{\pi M}{\alpha_i}\right) - \sin\left(\frac{\pi M}{\alpha_i}\right) \Big/ \frac{\pi M}{\alpha_i} \right] \mathrm{III}(M/\alpha_i) \qquad (80)$$

This function represents a set of points along the M axis at positions

$$(L, M) = (0, k\alpha_i) = (0, k \Delta\tau R T_i^{-1} \nu_0 \sec\delta_0) \qquad (k = \pm 1, \ldots) \qquad (81)$$

with amplitudes

$$g_{1i}(L, M) = \frac{\nu' T_i \cos\delta_0}{k R \nu_0} (-1)^k \qquad (k = \pm 1, \ldots) \qquad (82)$$

For continuum observations where you have to integrate Eq. (82) over the bandwidth, the sidelobes disappear since Eq. (82) is symmetric in ν'. However, if you observe a source with a very pronounced spectral distribution sidelobes as given in Eq. (82) will appear. For sources away from the fringe-stopping center, the effect described above will be in addition to the effect of delay-bandwidth effects, as described in the next section.

c. Finite Bandwidth. Even if the combination of fringe stopping and delay produces the correct frequency-dependent phase correction for an ideal coordinate shift, the map obtained will have errors depending, mainly, on the width of the band filter. Physically, you can consider this as due to the fact that a source away from the fringe-stopping center will have an additional, uncorrected time delay. This time delay will produce a modulation in the frequency domain. The integration over a finite frequency band, due to the fact that we sample the correlation function at one time difference $t = 0$ only (cf. Section II), will diminish the response on the map as soon as the additional time delay is no longer much smaller than the inverse of the band filter width.

To obtain an idea of the influence of this effect, we will look at it in a different way. The map obtained by taking the Fourier transform of the observed correlation function is, in essence, the integral of all the maps at different frequencies. These maps have scales depending on the frequency; however, the sidelobe structure will have the same scale. Writing $\beta(\nu')$ for the bandpass function, where $\nu' = \nu - \nu_0$, we have

$$B_{\text{obs}}(\rho, \phi) = \frac{\int_{-\infty}^{+\infty} B\left(\rho - \frac{\nu'}{\nu_0}\rho, \phi\right) \beta(\nu') \, d\nu'}{\int_{-\infty}^{+\infty} \beta(\nu') \, d\nu'} \tag{83}$$

This will broaden the synthesis beam pattern, on the one hand, and, on the other hand, reduce the intensity. The effect can be described as a radial convolution with a function $\beta(\rho \, \nu'/\nu_0)$.

d. Finite Sampling Interval. In practice the observations are averaged before they are recorded. The influence of this averaging can be very easily seen in the case where it is a straightforward integration, and the integration interval on the (u, v) plane is less than the distance between successive ellipses on the plane. In that case the integration can be looked at as a convolution with a hatbox function, and the influence on the map is a straight multiplication with a circular $J_1(a\rho)/a\rho$ function. The width is inversely proportional to the width of the integration interval. If the integration interval is different for all ellipses in such a way that the arc traversed is the same in the (u, v) plane for all ellipses, the resulting map will be simply multiplied by the above function. In practice the integration time is often the same for all ellipses, and hence, the multiplying function is different for the different baselines. The resulting position-dependent synthesized beam pattern is than the sum of the $J_0(x)$ functions composing the total beampattern (cf. Section III, C, 4), weighted according to the value of the multiplying function.

If the time constant function is not a simple symmetric function, the effect on the map will be equal to a convolution with the time constant function. At any one place this convolution must occur perpendicular to the radius vector connecting the point on the map with the center of the map. The characteristic width of the function will be proportional to the distance from the field center (cf. Brouw, 1971).

There is one other effect that also plays a role, i.e., the finite sampling distance between points along the ellipses. This sampling can be described by the multiplication of the correlation function with a function consisting of a set of radial spikes. The influence on the map will be a convolution with a function consisting of a set of radial spikes again. The net effect will be that the resultant antenna pattern will be the sum of a set of antenna patterns convolved with straight lines, or, if we consider only the main beam of the pattern, the sum of a set of line patterns, each one constant along a line, but perpendicular to the line with the same pattern as a cut through the original beam pattern. As long as the distance between the lines is small compared to the beamwidth the resultant pattern will be roughly equal to the original pattern. However, at distances from the field center where the distance between lines becomes of the same order as the beamwidth, a set of radial sidelobes will appear.

e. Primary Beam. The antenna pattern of a paraboloid can, at least for the central region, be represented by

$$P(\rho, \phi, \nu) = \left[\frac{2J_1(\alpha \nu \rho)}{\alpha \nu \rho}\right]^2 \qquad (84)$$

where α is a fitting constant. Writing again $\nu' = \nu - \nu_0$, for $\nu'/\nu_0 \ll 1$ we have

$$P(\rho, \phi, \nu) = P(\rho, \phi, \nu_0)[1 - 4\nu'/\nu_0 + 2\alpha \nu' \rho J_0(\alpha \nu_0 \rho)/J_1(\alpha \nu_0 \rho)] \qquad (85)$$
$$= P(\rho, \phi, \nu_0)[1 - \beta(\nu'/\nu_0)]$$

The factor $1 - \beta(\nu'/\nu_0)$ can be considered as a bandwidth effect (cf. Section III, C, 4, c), and the influence determined in the same way. This is, of course, as should be expected, since the frequency-dependent beamwidth will change the apparent spectrum of the source.

f. Noise. The Fourier transform is a linear process and, hence, pure noise on the observed correlation function will produce noise on the map.

Phase noise is another problem again. Its main effect, in addition to producing noise, can be described as producing a reduction of the real part of the antenna pattern. For Gaussian noise the average reduction will be equal to $1 - \exp[-\sigma^2/2]$, where σ^2 is the variance of the distribution. For $\sigma = 5°$ this amounts to about 3.8‰ and for $\sigma = 10°$ to 1.5%. These last figures are only true for uncorrelated phase noise from point to point in the (u, v) plane. In practice, this will not be true. Phase fluctuations caused by irregularities in the atmosphere, are, for instance, often correlated for different simultaneously observed baselines.

D. Sensitivity

The sensitivity of an aperture synthesis telescope expressed as the rms error σ in the measured flux density of a point source is given by (Högbom and Brouw, 1974)

$$\sigma_S = \frac{M}{\eta} \frac{k\,T_{\text{sys}}}{(\Delta\nu\,t)^{1/2} A_{\max} P(\rho, \phi) N^{1/2}} \quad (\text{W m}^{-2}\,\text{Hz}^{-1}) \qquad (86)$$

where $k = 1.38 \times 10^{-23}$ W Hz^{-1} deg^{-1} (Boltzmann's constant), T_{sys} = system noise temperature (in °K), M = factor expressing the degradation of the particular receiver as compared to an ideal square law detector (cf. Christiansen and Högbom, 1969) (this factor is in most practical cases between 1 and 2), η = degradation in sensitivity due to unequal weighting of the different interferometers (η is not very sensitive to the exact shape of the weighting function, and is of the order of 0.9), $\Delta\nu$ = receiver equivalent frequency bandwidth (in Hz), t = total observing time (sec), A_{max} = effective area of an individual interferometer in the maximum direction (m²) (it equals the geometric mean of the geometric areas of the two interferometers elements, multiplied by the aperture efficiency), P = normalized power pattern of the interferometer primary beam, N = number of simultaneously operating interferometers.

Equation (86) is valid under the assumption that all interferometers are identical, and all interferometers have uncorrelated noise. The measured flux and its error will then refer to the component of the incoming radiation whose polarization is "matched" to that of the antennas. This rms error is not equal to the sensitivity of the instrument. The sensitivity will depend on the definition of the problem in hand. For instance, a 2σ deflection at a well-defined position on the map, e.g., while looking for a radio source at the position of an optical source, will have a probability of 97.7% of being caused by a radio source. However, in a full map of 10^5 points, one can expect 3 points with a deflection of 4σ, or more, due to noise.

The sensitivity for extended objects depends on the maximum and minimum interferometer spacing, the weighting function used, and the equivalent solid angle of the synthesized beam.

The relation between a deflection on the map of S W m^{-2} Hz^{-1} and the corresponding brightness temperature T_{obs} is given in the case of an East–West interferometer by

$$T_{obs} = \pi S k^{-1} R_{max}^2 \sin \delta_0 \int_0^1 r\, g(r)\, dr \qquad (87)$$

where R_{max} = maximum baseline (m), $r = (u^2 + v^2)^{1/2}/R_{max}$, δ_0 = declination of field center, $\sin \delta_0$ = factor reflecting the equivalent synthesized beam solid angle, $g(r)$ = normalized weighting ("grading") function [$g(0) = 1$].

This equation only holds for sources small in angular measure compared to the radius of the first grating ring and small compared to the broad negative sidelobe originating from the missing short baseline interferometers. For large sources one should get rid of both the grating rings and the extensive negative sidelobe by observing at enough interferometer baselines, on the one hand, and inserting the small spacing information by, for instance, observing them with a large single element telescope. If these rules cannot be followed, one should go back to the initial equations, and the relation between the deflection on the map expressed as a flux density (S), and the brightness temperature (T_b) is given by

$$S = \frac{2k}{\lambda^2} \int_{\text{source}} T_b G \, d\Omega \tag{88}$$

where G = normalized synthesized beam pattern $[G(0) = 1]$ and the integration extends over the full solid angle of the source. The sensitivity to weak extended sources can be improved by smoothing the map to correspond to observations that have been made with a wider synthesized beam. Simple considerations would lead one to expect an improvement of $(\Omega_1/\Omega_0)^{1/2}$, where Ω is the solid angle of the synthesized beam. However, the improvement is larger and equal to a factor $(\Omega_1/\Omega_0)^{3/4}$. This is due to the fact that smoothing widens the synthesized beam in both dimensions with a factor of say, q, which, in effect, means that we are using measurements from an array $1/q$ times as long. The number of interferometers N in Eq. (86) and the maximum baseline $R_{\max}$ in Eq. (87) are both reduced by a factor q. The result is that σ_s increases by a factor $q^{1/2}$ (or $(\Omega_1/\Omega_0)^{1/4}$), but σ_T decreases by $q^{3/2}$ (or $(\Omega_1/\Omega_0)^{3/4}$).

From the foregoing it is clear that smoothing to a wider beamwidth, or, which is equivalent, using a narrower grading function, has a relatively small influence on the sensitivity for point sources, and a much larger influence on the brightness sensitivity.

It is often claimed that combining all data points with equal weight gives the best sensitivity as far as point source detection goes. Especially, it gives a much better sensitivity if compared to the normal practice of weighting each point on a circle (in the case of an EW interferometer) with the radius of the circle. However, this advantage is only a factor $\sqrt{\frac{3}{2}}$ in the limit for an infinite number of rings (Thompson and Bracewell, 1974), and only in those cases were the sidelobe level is completely irrelevant is it worth the added disadvantage of higher sidelobe level. The factor $\sqrt{\frac{3}{2}}$ can easily be deduced if we consider N rings and weight the nth ring with a factor n. In that case the sensitivity for a point source is lower than in the case of uniform weighting by a factor

$$\left[\left(\sum_{n=1}^{N} n^2\right)^{1/2} \bigg/ \sum_{n=1}^{N} n\right] \frac{N}{N^{1/2}} = \left[\frac{2(2n+1)}{3(n+1)}\right]^{1/2} \tag{89}$$

It should be noted that Eqs. (86)–(87) are based on the assumption that all points of the measured correlation function have independent noise.

Therefore, in practice the rms error will be higher. The main difference is that the error will not be inversely proportional with the square root of the number of interferometers but proportional with $N^{-\alpha}$, where α is less than 0.5. Furthermore, in practice the sensitivity will be lower due to calibration errors, and nonperfect behavior of the instrument.

1. Dynamic Range

The sensitivity as given in Section III, D is only valid if no sources are present on the sky, or, if the behavior of the instrument is fully known. In practice, however, the response of the instrument for point sources of radio radiation, i.e., the synthesized beam pattern, will be known with a limited accuracy only. In Section III, C, 4 some of the sources of error are given, but many more can be thought of. These discrepancies between the expected and real beam pattern will add to the noise in the map, since the interpretation of a map will be based on an assumed beam pattern. Let us consider an example of a completely filled (u, v) plane, producing a theoretical beam pattern $2J_1(a\,\rho)/(a\,\rho)$.

This pattern has a strong main peak with a width between the first zeros of about $7.7\,a^{-1}$ in ρ. The sidelobes are at positions

$$\rho_k \approx (\tfrac{3}{4}\pi + \tfrac{1}{2}\pi\,k)\,a^{-1} \qquad (k = 0, 1, \ldots) \tag{90}$$

and have intensities

$$j_{l,k} \approx (-1)^k \left(\frac{8}{\pi}\right)^{1/2} (a\,\rho_k)^{-3/2} \qquad (k = 0, 1, \ldots)$$

or for large k

$$j_{l,k} \approx (-1)^k \frac{8}{\pi^2} k^{-3/2} \qquad (k = 0, 1, \ldots) \tag{91}$$

We have a point source on a map, for which we have determined the position and intensity. To get rid of the sidelobes produced by this point source, we subtract, one way or another, the response of this point source from the map. However, let us assume we subtract the source at a slightly wrong position, say $\Delta\rho$. This will produce a difference pattern in the direction of the shift, equal to

$$\Delta j_{l,k} \approx [2J_2(a\,\rho)/a\,\rho]\,\Delta(a\,\rho) \tag{92}$$

This function has sidelobes at

$$\rho_{2,k} \approx (\tfrac{5}{4}\pi + \tfrac{1}{2}\pi\,k)\,a^{-1} \qquad (k = 0, 1, \ldots) \tag{93}$$

with intensities, for large k

$$\Delta j_{l,k} \approx (-1)^k \frac{8}{\pi^2} k^{-3/2} \Delta(a\, \rho_{2,k}) \qquad (k = 0, 1, \ldots) \qquad (94)$$

It will be very difficult to distinguish this pattern from the noise, especially if there are more difference patterns from more sources. Difference patterns also occur, of course, for all other types of errors—beam broadening, map scale errors, phase errors, etc. In practice, the addition of all these difference patterns is, especially if one or more strong sources are present in, or close to, the map quite often the limitation in sensitivity. Detection of sources below a level of about 0.004 of the strongest source in a map is very difficult indeed, and below 0.002 asks for precision up to now, as far as I know, not obtained. In special positions of the map a higher dynamic range can be obtained. The actual dynamic range is a function of the sidelobe level. Hence, near areas of low sidelobe level a higher precision can be attained.

2. Confusion

In radio astronomy the term "confusion" is used to refer to the fact that every observed field contains a large number of weak sources. These cause deflections that merge to a noiselike distribution over the map. The confusion problem is to decide how many deflections on a map can be interpreted as due to individual sources rather than to a blend of many weaker sources. The often stated rule in radio astronomy is to accept the largest deflections as individual sources, but not to accept more sources than corresponds to an average of one source per 20–30 beam areas. So, even if an instrument has a sensitivity of, say, 10^{-29} W m^{-2} Hz^{-1}, the useful sensitivity may be limited to 10^{-28} W m^{-2} Hz^{-1}. The aperture synthesis telescopes operational today have no serious confusion problems at their normal operating frequencies ($\sim$1–5 GHz). However, at frequencies below a few hundred megahertz this problem can occur, although for higher frequencies problems may already be present, especially for surveys down to a low intensity limit.

The confusion problem is even more complicated in the presence of grating rings and other prominent sidelobe patterns. Burns (1972) has discussed the statistical theory of the confusion problem and the influence of the detailed shape of the reception pattern.

E. Single Beam vs. Aperture Synthesis Telescope

Aperture synthesis telescopes have two main advantages as compared to most other types of radio telescope. First, a very high resolving power, for the amount of antenna structure actually built, can be achieved. Second, an aperture synthesis telescope measures directly a radio image of a complete field, rather than the radio brightness in one single direction only. A single-beam instrument with the same collecting area and receiver quality, e.g., a Cross antenna (e.g., Christiansen and Högbom, 1969), must spend the same time on each of the directions in the map separately in order to produce the same map with the same sensitivity. Hence, to observe individual sources the two types of telescope have identical performance. However, for mapping purposes the aperture synthesis telescope is far superior. It is, of course, completely out of the question to build single dish telescopes with the same resolving power as that of the existing aperture synthesis telescopes, which have dimensions of the order of one up to a few kilometers. If one could build such a dish, how would its performance compare with a synthesis telescope? Let us assume a synthesis telescope with N identical interferometers consisting of elements with a diameter d, a geometrical area $(\pi/4)\,d^2$, and a maximum baseline R. The beams of the two instruments are equal, if the large dish has a geometrical area $(\pi/4)\,R^2$. Let us, furthermore, assume that we need M periods of 12 hours to obtain a grating ring free map of the area of sky within the halfpower beam of the interferometer element. This halfpower beam contains $(R/d)^2$ synthesized beam areas. In M periods the sensitivity of the synthesis telescope is equal to (Section III, D)

$$\sigma_S = c\,(M\,N\,t)^{-1/2}\,d^{-2} \qquad (\text{W m}^{-2}\,\text{Hz}^{-1}) \tag{95}$$

where c is a telescope and receiver dependent constant.

The single dish sensitivity with the same type and quality receiver and telescope will be

$$\sigma_D = c\,t^{-1/2}\,R^{-2} \qquad (\text{W m}^{-2}\,\text{Hz}^{-1}) \tag{96}$$

For a single point source the dish can obtain the same sensitivity as the aperture synthesis system in a fraction of the time only.

$$t_D = (d/R)^4\,M\,N\,t_s \tag{97}$$

A practical case is a telescope where $d = 25$ meters, $R = 1600$ meters, $N = 40$, $M = 4$, $t_s = 12$ hours. In that case $t_D = 0.4$ sec, a time only 2×10^{-6} of the time needed by the synthesis telescope (4×12 hours). However, for mapping, the superiority is much less pronounced. The single dish has to observe $(R/r)^2$ beam areas. For full coverage 4 points per beam area are necessary. Hence, the total time spent on mapping an area with the same sensitivity as the synthesis telescope, neglecting the time needed to move from one point to the other, is equal to

$$T_D = 4(d/R)^2\,M\,N\,t_s \tag{98}$$

For the telescope mentioned above, this amounts to 1.9 hr, almost 4% of the time needed by the synthesis telescope, for a cost, if it could be built, of $\frac{1}{12}$ $(R/d)^{2.5}$ that of the synthesis telescope. The $\frac{1}{12}$ arises from the fact that 12 telescopes were necessary to build the above mentioned telescope, the power 2.5 is an experience number for the normal range of telescope diameters. Hence the single dish will work 25 times as fast for a 3000 higher expense, if it could be built at all. In practice multiple beams could be made for a single dish. In that case the range between the speed and cost is less pronounced, and the receiver complexity is more comparable between the two systems. However, it will remain cheaper to build a number of identical synthesis telescopes if one wants to have the single-dish observing speed.

IV. Data Processing

To observe the spatial correlation function and to convert it into a map of the radio brightness distribution of the sky requires a lot of computing power. In principle one would like to handle the data with an analog device. This would be the most obvious and most practical way. However, attempts to build a multidimensional analog device (e.g., McLean et al., 1967) have not been able to reach acceptable tolerances as yet. However, a one-dimensional analog device, built to obtain the spectral information, has been constructed with success (Cole, 1973). We are, therefore, restricted to the use of digital computers, either general or special purpose, to handle the aperture synthesis telescope data. We can split the necessary computations into several groups:

A. Telescope steering and data collection
B. Calibration
C. Correction
D. Transformation of correlation function into brightness distribution
E. Displaying and handling the resultant map of the sky.

In practice these functions can overlap in several ways, but in about all cases a division as indicated above suggests itself logically.

A. Telescope Steering and Data Collection

The steering of a synthesis telescope consists of two parts. On the one hand, the interferometer elements must be pointed to the correct place in the sky. This pointing is the same as if it were done for a single element telescope. Hence, a pointing program should follow the diurnal rotation of the sky and correct the pointing coordinates for the known pointing corrections due to structural deformations of the element, and for the refraction of the incoming radio waves by the Earth's atmosphere.

On the other hand, the correlation receiver connecting the interferometer elements should be steered. This steering must include the setting of the delays and the running of the fringe-stopping mechanism. It could also include the correction of the baselines and positions to be used in calculating the values for the delay and fringe-stopping settings for known variations, as there are the precessional variations of the coordinates, the nonuniform rotation of the Earth, the variation of the baseline position with respect to the axis of rotation of the Earth due to the polar motion, known variations in the apparent baselines due to differential structural errors of the elements, and so on. However, if these corrections are sufficiently small, as is usually the case in practice, they can also be applied as corrections to the observed correlation function in a subsequent correction program.

The steering tasks are best done with a general purpose computer, since better known parameters for the telescope can easily be inserted. However, the use of a special purpose computer will adequately fulfill its purpose (O'Sullivan *et al.*, 1973).

Data collection can be done either by sampling the correlation function and storing it for later reference, e.g., on magnetic tape or disk, or by making the samples available to another program for real-time computation. Most synthesis telescopes at present store the data rather than using it straight away. The notable exception being the Cambridge 5-km array (Ryle, 1972).

B. Calibration

The precision of the positions obtained with aperture synthesis depends directly on the precision with which the position of the interferometer baseline is known with respect to the sky. Surveyance with geodetic means cannot, in general, obtain a precision comparable to that which is astronomically obtainable. Furthermore, the position of the baseline with respect to the sky is dependent on the rotational parameters of the Earth. Normal (optical) observations done to determine the rotational parameters of the Earth cannot obtain an absolute precision in time better than to about 4 msec.

To obtain the best possible value of the baseline parameters, calibration with the aid of celestial radio sources is the obvious way. A radio point source with a known position and intensity will produce a correlation function that can be exactly calculated for given baselines. Any departure of the observed correlation function from the theoretical function will be due to errors in the assumed position of the source, errors in the assumed position of the baseline with respect to the sky, differential errors in the interferometer elements in different directions of the sky, or instabilities in the receiving equipment. The influence of position and baseline errors on the theoretical response can easily be deduced from the formulas given in Section III, A. The behavior of the other errors will depend on the actual structural properties of the interferometer elements and on the special way the receiving equipment is built. From the observed output of a set of radio point sources the instrumental errors can be determined (e.g., Wade, 1970; Ryle and Elsmore, 1973; O'Sullivan and Brouw, 1973; Brouw and Van Someren Gréve, 1973).

If an adequately chosen set of calibration sources is selected, the actual positions of the sources do not have to be known. With an aperture synthesis telescope the determination of declinations can be absolute, while for the determination of right ascension only a zero point and known rotational parameters of the Earth have to be assumed (Wade, 1970; Ryle and Elsmore, 1973).

Even in the case where the interest in precise positions is low, the knowledge of the behavior of the instrument is of great importance. The precision with which the output can be predicted determines the precision with which the synthesized beam pattern is known, and, hence, determines the dynamic range of the instrument.

Calibration observations are also necessary to determine the phase and gain errors of the receiving equipment. In some cases the behavior of the receiver is monitored by a calibration system (Hamaker, 1974; Frater et al., 1973), which, of course, can determine the receiver behavior on a much shorter time scale than with the aid of calibration observations only.

C. CORRECTION

The observed correlation function can be corrected for small errors in the telescope system after it has been observed. Some errors will be known, or known with a higher accuracy, only after the observations have been done. Examples are the rotational parameters of the Earth, atmospheric and ionospheric parameters, interpolation rather than extrapolation of calibration results.

The exact number of corrections depends, again, largely on the specific instrument used (e.g., Brouw and Sullivan, 1973; Brouw, 1971; Wade, 1970).

D. Transformation of Correlation Function into Brightness Distribution

As has been shown in Section III, A, the relation between the observed correlation function and the brightness distribution is only a two-dimensional Fourier transform in the case of an aperture synthesis telescope consisting of interferometers with East–West baselines only. In all other cases a three-dimensional Fourier transform, which is only defined on a nonplane surface in space, or a transform of the form

$$B(l, m) = \int\int_{-\infty}^{+\infty} P(U, V) \tag{99}$$
$$\times \exp[-2\pi i(U l + V m + W(1 - l^2 - m^2)^{1/2}] \, dU \, dV$$

is necessary. The correlation function will be sampled only in points in the (U, V, W) space, and, furthermore, will be weighted to obtain a beam pattern most suited for the purpose of the specific observations. Combining the sampling and weighting function in one grading function $G(U, V)$ we have

$$B_{\text{obs}}(l, m) = \int\int_{-\infty}^{+\infty} G(U, V) \, P(U, V) \tag{100}$$
$$\times \exp[-2\pi i(U l + V m + W(1 - l^2 - m^2)^{1/2}] \, dU \, dV$$

producing a map of the sky with a beam pattern $g(l, m)$

$$g(l, m) = \int\int_{-\infty}^{+\infty} G(U, V) \tag{101}$$
$$\times \exp[-2\pi i(U l + V m + W(1 - l^2 - m^2)^{1/2}] \, dU \, dV$$

The direct transform of Eq. (100) is the most straightforward way to obtain the estimate for the brightness distribution. However, to produce a map of N points from M correlation samples requires NM operations, each operation consisting of a complex multiplication and addition. With N and M both of the order of a hundred thousand, the computing involved limits the possibilities severely. This method has the advantage that during the transform any correction depending on both the position of the source and the position of the correlation sample, e.g., bandwidth effects, differential refraction across the primary beam, can easily be applied, while the non-East–West interferometer is easily accommodated. In practice a square map of $N_m \times N_m$ points is often required, the number of correlation samples in general being of the same order. Hence, $NM \simeq N_m^4$. With the value of N_m involved, it is worthwhile looking for methods with a computing time depending on a lower power of N_m. In the case of a line array synthesis telescope, one could perform a fast Fourier transform (Cooley and Tukey, 1965) on the line of correlation samples obtained at each sampling time. In that way the computation expense can be brought down to a value proportional to $N_m{}^3$.

Both this last method and the direct transform have the advantage that the brightness distribution can be built up while the observations are underway (Kenderdine, 1974). Corrections depending on later calibrations, however, have to be applied via a convolution in the map.

Other existing methods include interpolation along the samples along a radial line for the case of an East–West interferometer (Thompson and Bracewell, 1974), and a method based on the splitting of the (U, V) plane into quadrants (Kenderdine, 1974). However, to reduce the computing time to a value proportional to about $N_m{}^2$, full use must be made of the fast Fourier transform.

1. Fast Fourier Transform

The direct method of Eq. (100) can, in the case of existing large synthesis instruments, only be used if either the number of observed correlation samples or the number of brightness points wanted is small. A small number of correlation samples can be obtained if a field on the sky is not observed continuously, but only at specific times to produce enough samples to enable the determination of flux, position, and extension parameters. A small map is a much less easy proposition. It is only feasible if it is well known that the part of the sky under observation contains only a few sources, much brighter than the remaining ones in the field of view of the telescope elements, or if the properties of the other sources are well known. In all other cases sidelobes from sources outside the small area of direct interest will uncorrectably interfere with the brightness distribution of primary interest. A fast Fourier transform (FFT) algorithm (Cooley and Tukey, 1965) can help in reducing the computing load to a more acceptable $N_m{}^2$ number of operations. In the following, only the two-dimensional (East–West) case will be treated. The non-East–West case can always be treated as a set of two-dimensional transforms, which must be connected by a transform in the third dimension (cf. Section III, A). This last transform can be a direct transform, since only one point in the third dimension is of interest. The FFT cannot be used if corrections that depend on both the position of the correlation function sample and the position of a brightness point have to be applied. Hence, the design tolerances of a synthesis telescope should be constrained by the requirement to keep the value of the necessary corrections of this type below a minimum level, in such a way that the influence of the errors introduced by not correcting for the effects is less than the general precision of the result. The FFT can only be used on sample points arranged on a rectangular grid. The observed samples lie, in general, on parts of ellipses. To convert the observed sampling grid into a rectangular grid some kind of "gridding" process is necessary.

a. Influence of Rectangular Grid. An FFT with n_u and n_v points in, respectively, the U ($\rightarrow l$) and $V(\rightarrow m)$ coordinate and a spacing ΔU and ΔV between the grid points will produce a map of the sky of an area with widths $(\Delta U)^{-1}$ and $(\Delta V)^{-1}$ in, respectively, the l and m directions. The spacing between the brightness samples will be equal to $[n_u(\Delta U)]^{-1}$ and $[n_v(\Delta V)]^{-1}$.

The rectangular grid will produce sidelobes on the map, in addition to the sidelobes already caused by the observational gridding. The multiplication of the observed correlation function with a "nail-bed" function causes the brightness map to be convolved with a nail-bed function. This last function can be represented by

$$N(l, m) = \delta[l - (\Delta U)^{-1}] \, \delta[m - (\Delta V)^{-1}] \quad (102)$$

where $\delta(x)$ is the Dirac δ function. The effect of this convolution is that "aliasing" will occur, i.e., the brightness map will be repeated at intervals $(\Delta U)^{-1}$, respectively, $(\Delta V)^{-1}$ in the l and m directions. The map obtained will, hence, be equal to

$$B_{\text{obs}}(l, m) = \sum_{k=-\infty}^{+\infty} \sum_{j=-\infty}^{+\infty} B[l + k\,(\Delta U)^{-1}, m + j\,(\Delta V)^{-1}] \quad (103)$$

That is, features from outside the transformed area will appear in the map, although, due to the primary beam of the interferometer elements and the effects of delays and fringe stopping, with reduced intensities. The "gridding" procedure used to obtain the rectangular grid should be designed in such a way as to minimize the effects of this "aliasing."

b. Gridding Procedures. The best possible gridding process, from the viewpoint of the astronomer going to use the map, would be to convolve the observed correlation function with

$$G(U, V) = \frac{\sin(\pi U/\Delta U)}{\pi U/\Delta U} \frac{\sin(\pi V/\Delta V)}{\pi V/\Delta V} \quad (104)$$

This has the effect of multiplying the produced map with a two-dimensional rectangular pulse covering the transformed area exactly. Hence, the aliasing process as described by Eq. (103) will have no influence on the final result. However, the convolution process will require N_m^4 operations, and no gain in computer time is obtained as compared to the direct transform process. The simplest gridding process is "cell summing"

(Mathur, 1969). All observed points within a rectangle with sides ΔU and ΔV centered on a rectangular grid point are added together and this sum assigned to the rectangular grid point. The effect will be to produce a map

$$B_{\text{obs}}(l, m) = \sum_{k=-\infty}^{+\infty} \sum_{j=-\infty}^{+\infty} B(l + k\,\Delta U^{-1}, m + j\,\Delta V^{-1}) \frac{\sin[\pi(\Delta U\,l + k)]}{\pi(\Delta U\,l + k)} \\ \times \frac{\sin[\pi(\Delta V\,m + j)]}{\pi(\Delta V\,m + j)} \tag{105}$$

Hence, aliasing will occur, although for the center of the map all aliased components will cancel each other. The main drawback of this method, however, is that the resultant beam pattern is not shift independent (Thompson and Bracewell, 1974); that is, point sources in different places of the map will have different synthesized beam patterns. A close relative of this method called cell averaging is used with the National Radio Astronomy Observatory (NRAO) interferometer (Hogg et al., 1969). In this case the average, rather than the sum, of all observed correlation function samples within a rectangle centered on a rectangular grid point is assigned to the rectangular grid point. The Westerbork synthesis telescope uses a Gaussian convolution function to obtain the rectangular grid points (Brouw, 1971). It covers an area of about 15 cells on the (U, V) plane and is dimensioned in such a way that the multiplying function on the brightness maps has a value of 0.25 on the axes at the edges of the map as compared to the center value. The resulting brightness map is then equal to

$$B_{\text{obs}}(l, m) = \sum_{k=-\infty}^{+\infty} \sum_{j=-\infty}^{+\infty} B(l + k\,\Delta U^{-1}, m + j\,\Delta V^{-1}) \\ \times \exp[-8 \ln 2(l\,\Delta U^2 + m\,\Delta V^2)] \tag{106}$$

The resultant tapering can be corrected for by a simple multiplication, and aliasing is mostly restricted to a band at the edges of the observed field. The computations are, of course, more time consuming that for simple cell summing or averaging.

Thompson and Bracewell (1974) give an extensive description of the relative merits and properties of the different gridding procedures described above. There remains, however, the question of whether an optimum gridding procedure can be defined.

c. *Optimum Gridding.* The optimum gridding is obtained if we convolve the observed correlation function with the function defined by Eq. (104). However, as has been shown in Section IV, D, 1, b, the computing time involved is prohibitive. Let us, therefore, define our optimum requirements in view of the computing time we are willing to spend on the gridding process. In other words, if we are willing to use a convolution function with an extent of $n \times m$ rectangular points, what is the optimum function to be used? Again, an optimum function may depend on the end result of our transform process. In some cases, for instance, a Gaussian function may be the optimum, since it leaves the central area of the map alias-free, and by producing a large enough map the central area may have any size. However, the computing time involved in producing too large a map, especially in the larger synthesis systems coming into use, will prohibit this waste of a part of the map. Let us try to define an optimum function giving the "least overall aliasing" possible.

Let us define our convolution function $F(U, V)$ as being zero outside the area $(-n/2 \Delta U, -m/2 \Delta V) - (n/2 \Delta U, m/2 \Delta V)$. This function will have a Fourier transform $f(l, m)$. The uncertainty principle states that the concentration of $f(l, m)$ is limited and depends on the concentration of $F(U, V)$. Let us define the mean-square concentration of $f(l, m)$ in the map area as

$$\alpha^2 = \frac{\int_{-(2\Delta U)^{-1}}^{(2\Delta U)^{-1}} \int_{-(2\Delta V)^{-1}}^{(2\Delta V)^{-1}} f^2(l, m) \, dl \, dm}{\int_{-\infty}^{+\infty} \int_{-\infty}^{+\infty} f^2(l, m) \, dl \, dm} \tag{107}$$

that is, α^2 is a measure of the "energy" of $f(l, m)$ contained in the transformed map area, as compared to the total energy of $f(l, m)$.

If we could find a function $F(U, V)$ that would maximize α^2, we will have found an optimum convolving function, which minimizes the amount of energy going into the aliasing. Whether the energy requirement is always the optimum requirement remains, of course, open to dscussion. I do not know whether a two-dimensional solution of this equation exists, but we can always look at the product of two one-dimensional function $f(l), f(m)$ and ask the same question. Maybe this is not the optimum solution, but for computing reasons it is certainly a more attractive one. Hence, we ask the question, can we maximize α^2 as given by

$$\alpha^2 = \frac{\int_{-(2\Delta U)^{-1}}^{(2\Delta U)^{-1}} f(l)^2 \, dl}{\int_{-\infty}^{+\infty} f(l)^2 \, dl} \tag{108}$$

and since, without loss of generality, we can normalize $f(l)$ in such a way that

$$\int_{-\infty}^{+\infty} f(l)^2 \, dl = 1 \tag{109}$$

as given by

$$\alpha^2 = \int_{-(2\Delta U)^{-1}}^{(2\Delta U)^{-1}} f^2(l) \, dl \tag{110}$$

The optimum function $f_0(l)$ satisfies the integral equation (Landau and Pollak, 1961)

$$\lambda f_0(l) = \int_{-(2\Delta U)^{-1}}^{(2\Delta U)^{-1}} f_0(l') \frac{\sin[\pi n \Delta U(l - l')]}{\pi(l - l')} dl' \tag{111}$$

The maximum value of α^2 is reached for the largest eigenvalue λ_0 at which a solution exist. The function $f_0(l)$ is called a prolate spheroidal function (Slepian and Pollak, 1961). Using the notation of Flammer (1957), we have

$$\begin{aligned} f_0(l) &\propto S_{00}(c,\, 2l\,\Delta U) \\ \alpha_{\max}^2 = \lambda_0(c) &= (2c/\pi)[R_{00}^{(1)}(c,\, l)]^2 \\ c &= \tfrac{1}{2}\pi\, n \end{aligned} \tag{112}$$

The Fourier transform $F_0(U)$ of $f_0(l)$ is given by

$$\begin{aligned} F_0(U) &\propto S_{00}(c,\, 2U\,\Delta U^{-1}) & |U| &\leq \tfrac{1}{2}n\,\Delta U \\ &= 0 & |U| &> \tfrac{1}{2}n\,\Delta U \\ c &= \tfrac{1}{2}\pi\, n \end{aligned} \tag{113}$$

The angular (S_{00}) and radial ($R_{00}^{(1)}$) prolate spheroidal functions are not expressable as well-studied functions. Rules for calculating them can be found, for instance, in Abramowitz (1965). For $n = 8/\pi \simeq 2.5$, λ_0 is already .9959 (Slepian and Pollak, 1961). From Eqs. (112)–(113) it can be seen that $\alpha_{\max}$ and f_0 depend only on the width of the convolving function, a result that was to be expected.

2. Other Transform Methods

Several attempts have been made to construct a method to transform the observed correlation function into a brightness distribution in a way differing from a straightforward Fourier transform. The purpose of these efforts was to produce a map with a synthesized beam pattern with lower sidelobes and higher resolution. Especially in those cases where the sampling of the correlation function is done in a nonregular way, the resulting sidelobes are often very cumbersome and limit the interpretation of the map considerably. Some of these methods were based on *a posteriori* adjustment of the map obtained with a standard Fourier transform (cf. Section IV, E). However, some were based on an interpolation process in the (U, V) plane to rid the observed correlation function of the zeros in those places where the correlation function has not been sampled, and on an extrapolation process to try to extend the area in which the correlation function was observed (e.g., Biraud, 1969; Burns and Yao, 1970).

These methods work only for observations with a high signal-to-noise ratio, or when assumptions about the map to be obtained can be made.

A notable exception is the maximum entropy method (MEM) to which attention has been drawn by Ables (1974). This method is also known in statistics as autoregression. The method has been proved to be a very valuable tool in the one-dimensional case of stripscans across a radio source or with an autocorrelation spectrometer (Ables, 1974; Lacoss, 1971).

The two-dimensional case is, at the moment at least, too expensive in computing time to be practically useful. However, if it becomes possible to reduce the computational needs drastically, the method should certainly be used, preferably in addition to standard methods, in aperture synthesis systems. The possibilities of obtaining a higher resolving power and lower sidelobes seem rather impressive.

The MEM is based on the following principle. There exist an infinite set of brightness functions whose Fourier transforms are all equal to the observed correlation functions. The crux of the matter is to select one of these functions as a reliable estimate of the true brightness distribution. The standard Fourier transform method selects the one which represents the observed correlation function in all points observed and transforms into zeros in all (U, V) points not sampled. This selection has the obvious advantage that it is an easily described process, and, hence, produces a result which is convolved with an exactly predictable beam pattern.

The MEM selects that member of the set of possible brightness functions which contains the minimum amount of information. A perfect example of the principle of Occam's razor. The difficulty, of course, is how to define "the amount of information." Information conveys the idea of order. The more ordered an environment, the more information it will communicate to the observer. The more chaotic a situation, the less information it will carry. Hence, it seems a logical step to equate the principle of the minimum amount of information with the maximum amount of chaos, or with the maximum amount of entropy. Hence, the brightness distribution selected will maximize the function

$$\int_{-\infty}^{+\infty} \ln B(l, m) \, dl \, dm \qquad (114)$$

the entropy gain of our transform process, under the constraints that the Fourier transform of $B(l, m)$ should equal the observed correlation function in all points sampled.

E. Map Handling

An anonymous array of numbers on a magnetic tape, or even printed on a piece of paper, does not convey much information. The astronomer who wants to interprete the estimate of the radio sky brightness obtained, wants to "see" his results. Hence, a lot of effort goes into ways to communicate the information available in the map to the astronomer.

First of all, the astronomer wants to look at his results, and programs to display the data as, e.g., contour diagrams, ruled surface maps, density modulated displays ("radiophotographs") are a first need. In addition to these the astronomer would like to have the best possible estimate of position and intensity of sources on the map. Programs to fit the theoretical beam pattern with all its sidelobes and grating lobes to features on the map should, therefore, be available. If all sources on the map are of interest, a search procedure which tries to find all sources present is necessary. Such a search program is quite simple, as long as one is only interested in sources with intensities above about ten times the noise. Below this level ingenious methods are necessary to be able to reach a reasonable degree of completeness (e.g., Hoekema, 1970).

One should be able to subtract from the map strong sources whose sidelobes and grating rings distort the picture as presented by the transform program, if the positions and the flux of the sources are known. This subtraction can be done either by subtracting the scaled beam pattern from the specified position on the map or by subtracting the theoretical correlation function for these sources from the observed correlation function, and retransforming the thus obtained correlation function. The last method is the superior one, since all signs of the source, including aliased sidelobes, will disappear. However, it is also an expensive method in computation terms, and will not always be justifiable. There will be programs to add maps, to extract polarization information from a set of maps and many more.

The above is only a selection from the almost unlimited number of map handling programs existing in all places where aperture synthesis maps are interpreted. Each astronomer will need special programs to interprete his special observations in the best possible way. In no way is there any difference between an astronomer and any scientist who has to deal with a lot of numerical results. It would be impossible to give a description of all programs available in one form or another.

An exception should be made for the program which has become known under the name of "cleaning" (Högbom, 1974). The dream of any astronomer is to be able to deconvolve his observed map with the beam pattern. In principle, of course, this is impossible. However, if one makes a set of assumptions, the cleaning process will produce a deconvolved picture of the sky. The basic assumption is that the amount of information available in a map is limited; i.e. only a small fraction of the observed sky contains sources above the sensitivity limit of the map. There is no theoretical background to prove that this is a necessary assumption. However, experience shows that the method breaks down if this assumption is not true (Högbom, 1974). Another assumption could be that the sky contains only positive sources; this however, has never been used in practical cases up to now. It is not quite clear whether this is a basic assumption. At first thought it looks like one, which makes the use of the method impossible for polarization maps.

The method proceeds as follows. The sky brightness map is scanned for the largest intensity in absolute value. At the grid point where this value occurs, a point source is assumed, the theoretical beam pattern is scaled to the intensity value at the point, and the beam pattern thus obtained is subtracted from the map. The scaling of the beam pattern is not to the full value of the intensity, but only to a fraction. This is necessary for two reasons. First of all, if it represents a point source, this source will, in general, not lie exactly on a grid point in the map. Second, the point may be part of an extended source, and contributions of neighboring points will be included in the observed value. The map is scanned again, and the next point is subtracted out. This process is repeated till either the maximum value approaches the map noise or reaches a specified level.

The set of quasi point sources is, in general, rebuilt into a map and added to the remainder of the original map. The rebuilding process is mostly done with a beam pattern that has about the same width as the main beam but is completely devoid of sidelobes, e.g., a Gaussian-shaped beam. The resultant map gives a much easier interpretable map than the original one, especially if the original synthesized beam had many sidelobes, due to, for instance, a nonregular sampled correlation function or a very scarcely sampled correlation function.

The convergence rate of the cleaning process depends on the number of points per synthesized beam pattern, and on the gain (g), i.e., the fraction of the actual intensity found used in the subtraction process used. The intensity remaining after n iterations is roughly proportional to $(1-g)^n$. However, as soon as the errors due to the allocation of point sources to grid points only become comparable to the remaining intensities, convergence slows down. The intensity left after n iterations becomes roughly proportional to n^{-1}. The level at which the transition between the two convergence domains occurs is proportional to the number of points per synthesized beam (Ekers et al., 1973).

There seems to be an indication that if the cleaning is continued to well within the noise, and rebuilding with a δ-function beam is done, the resultant map approaches the map that would have been obtained with the MEM (T. W. Cole, private communication). To obtain good results with the cleaning process, the theoretical synthesized beam pattern should

TABLE I
EARTH ROTATION APERTURE SYNTHESIS TELESCOPES

Location	Elements	Element beam	Frequency (MHz)	Max baseline (meters)	Synthesized beam	References[a]
Cambridge,	2	100'	1400	730	47"	1
England	3	175'	400	1550	80"	2
		50'	1400		23"	
		14'	5000		7"	
	8	6'	5000	4560	2"	3
Greenbank,	3	39'	1400	2700	16"	4
West Virginia[b]		20'	2700		8"	
		7'	8100		3"	
Stanford, California	5	7'	11000	206	19"	5
Fleurs, Australia[c]	68	100'	1400	800	40"	6
Big Pine, California[d]	3	77'	600	1080	16"	
		33'	1400		7"	
		27'	1700		6"	
		9'	5000		2"	
Pentiction, British Columbia	4	105'	1400	600	60"	7
Ooty, India	4	—	330	3500	48" × 330"	8
Hat Creek, California	2	11'	20000	300	9"	9
Westerbork, The	12	83'	600	1600	55"	10
Netherlands		36'	1400		23"	
		11'	5000		7"	

[a] Key to references: 1. Baldwin et al. (1970). 2. Ryle (1962). 3. Ryle (1972). 4. Hogg et al. (1969). 5. Bracewell et al. (1973). 6. Christiansen (1973). 7. Roger et al. (1973). 8. Swarup and Bagri (1973). 9. Hills et al. (1973). 10. Baars et al. (1973).

[b] Recently a 14-meter dish has been included in the system which can be used to provide baselines of 11 and 35 km. The maximum resolution of this system is about 1.2 at 1400 MHz.

[c] This instrument consists of both an East–West and a North–South synthesis telescope, each containing 34 elements.

[d] This instrument has both East–West and North–South synthesis capabilities, and has an additional 40 meter dish.

be as close to reality as possible. To obtain the ultimate, it should reflect all known imperfections of the instrument and of nature, which, in practice, means a different beam pattern for each point on the map. In practice, computer requirements limit the number of beam patterns that can be used. Hence, the obtainable result depends again on the stability and perfection of the instrument. Almost all synthesis telescope users have been using the cleaning process, and the results are very good. Standard procedures as used with the Westerbork synthesis array obtain dynamic ranges of up to about 20 dB. With care dynamic ranges of about 25 dB have been reached (R. Harten, private communication).

V. Conclusion

Since the pioneering work by radioastronomers (Ryle and Neville, 1963), many Earth-rotation aperture synthesis telescopes have been built. A review of the properties of the larger ones is given in Table I.

The astronomical results of these telescopes have shown the power of aperture synthesis. Most astronomical results with these telescopes are on extragalactic radio sources, although, of course, results on other sources are abundant as well. Many reviews of these results exist (e.g., Kellermann and Pauliny-Toth, 1973), and it falls outside the scope of the present work to go into the astronomical details.

There exist many plans for the future. On the one hand, most aperture synthesis telescopes are including an extra dimension to their possibilities by adding spectral resolution to their high spatial resolution. On the other hand, the maximum frequency is ever-increasing, and plans for millimeter wave aperture synthesis telescopes are well underway. The aperture synthesis telescope with the longest baseline being built at the moment is the VLA (Very Large Array) of the National Radio Astronomy Observatory (1967), which will consist of 27 telescopes arranged in a Y-shape. The maximum baseline is 36 km and the minimum wavelength about 2 cm. Some thoughts are being given to design very long baseline aperture synthesis telescopes. The future of aperture synthesis seems a very bright one indeed.

References

ABLES, J. G. (1974). *Astron. & Astrophys. Suppl.* **15**, 383.
ABRAMOWITZ, M., and STEGUN, I. A. (1965). "Handbook of Mathematical Functions." Dover, New York.
ALTENHOFF, W. (1966). Inaugural-Dissertation, Westfälische-Wilhelms-Universität, Münster.
BAARS, J. W. M., VAN DER BRUGGE, J. F., CASSE, J. L., HAMAKER, J. P., SONDAAR, L. H., VISSER, J. J., and WELLINGTON, K. J. (1973). *Proc. IEEE* **61**, 1258.
BALDWIN, J. E., JENNINGS, J. E., SHAKESHAFT, J. R., WARNER, P. J., WILSON, D. M. A., and WRIGHT, M. C. H. (1970). *Mon. Not. Roy. Astron. Soc.* **150**, 253.
BIRAUD, Y. (1969). *Astron. & Astrophys.* **1**, 124.
BORN, M., and WOLF, E. (1964). "Principles of Optics." Pergamon, Oxford.
BRACEWELL, R. N. (1958). *Proc. IRE* **46**, 97.
BRACEWELL, R. N. (1965). "The Fourier Transform and its Applications," Chapter 13. McGraw-Hill, New York.
BRACEWELL, R. N., and THOMPSON, A. R. (1973). *Astrophys. J.* **182**, 77.
BRACEWELL, R. N., COLVIN, R. S., D'ADDARIO, L.R., GREBENKEMPER, C. J., PRICE, K. M., and THOMPSON, A. R. (1973). *Proc. IEEE* **61**, 1249.
BROUW, W. N. (1971). Ph.D. Dissertation, University of Leiden.

Brouw, W. N., and O'Sullivan, J. D. (1973). *Proc. Inst. Radio Electron. Eng. Aust.* **34**, 341.
Brouw, W. N., and Van Someren Gréve, H. W. (1973). *Neth. Found. Radioastron. Int. Tech. Rep.* **112**.
Burns, W. R. (1972). *Astron. & Astrophys.* **19**, 41.
Burns, W. R., and Yao, S. S. (1970). *Astron. & Astrophys.* **6**, 481.
Christiansen, W. N. (1973). *Proc. Inst. Radio Electron. Eng. Aust.* **34**, 302.
Christiansen, W. N., and Högbom, J. A. (1969). "Radiotelescopes," Cambridge Univ. Press, London and New York.
Christiansen, W. N., and Warburton, J. A. (1955). *Aust. J. Phys.* **8**, 474.
Cohen, M. H. (1973). *Proc. IEEE* **61**, 1192.
Cole, T. W. (1973). *Proc. IEEE* **61**, 1321.
Cooley, J. W., and Tukey, J. W. (1965). *Math. Comput.* **19**, 297.
Ekers, R. D., Miley, G. K., and Le Poole, R. S. (1973). *Neth. Found. Radioastron. Note* **127**.
Flammer, C. (1957). "Spheroidal Wave Functions," Stanford Univ. Press, Stanford, California.
Frater, R. H., O'Sullivan, J. D., and Daniels, G. R. (1973). *Proc. Inst. Radio Electron. Eng. Aust.* **34**, 326.
Hamaker, J. P. (1974). *Neth. Found. Radioastron. Int. Tech. Rep.* **123**.
Hemenway, P. D. (1974). Ph.D. Thesis, University of Virginia, Charlottesville.
Hills, R. E., Janssen, M. A., Thornton, D. D., and Welch, W. J. (1973). *Proc. IEEE* **61**, 1278.
Hinder, R., and Ryle, M. (1971). *Mon. Not. Roy. Astron. Soc.* **154**, 229.
Hoekema, T. (1970). *Neth. Found. Radioastron. Int. Tech. Rep.* **87**.
Högbom, J. A. (1974). *Astron. & Astrophys. Suppl.* **15**, 417.
Högbom, J. A., and Brouw, W. N. (1974). *Astron. & Astrophys.* **33**, 289.
Hogg, D. E., Macdonald, G. H., Conway, R. G., and Wade, C. M. (1969). *Astron. J.* **74**, 1206.
Kellermann, K. I., and Pauliny-Toth, I. I. K. (1973). *Proc. IEEE* **61**, 1174.
Kenderdine, S. (1974). *Astron. & Astrophys. Suppl.* **15**, 413.
Ko, H. C. (1967). *IEEE Trans. Antennas Propagat.* **15**, 188.
Lacoss, R. T. (1971). *Geophysics* **36**, 661.
Landau, H. J., and Pollak, H. O. (1961). *Bell Syst. Tech. J.* **40**, 65.
McLean, D. J., Lambert, L. B., Arm, M., and Stark, H. (1967). *Proc. Inst. Radio Electron. Eng. Aust.* **28**, 375.
Mathur, N. C. (1969). *Radio Sci.* **4**, 235.
National Radio Astronomy Observatory. (1967). "A Proposal for a Very Large Array Radio Telescope," NRAO, Greenbank, West Virginia.
O'Sullivan, J. D., and Brouw, W. N. (1973). *Proc. Inst. Radio Electron. Eng. Aust.* **34**, 347.
O'Sullivan, J. D., Frater, R. H., Imrie, K. S., and Casse, J. L. (1973). *Proc. Inst. Radio Electron. Eng. Aust.* **34**, 319.
Roger, R. S., Costain, C. H., Lacey, J. D., Landecker, T. L., and Bowers, F. K. (1973). *Proc. IEEE* **61**, 1270.
Roger, A. E. E. (1968). *Tech. Rep.* **441**. Lincoln Laboratory, MIT, Cambridge, Massachusetts.
Ryle, M. (1962). *Nature (London)* **194**, 517.
Ryle, M. (1972). *Nature (London)* **239**, 435.
Ryle, M., and Elsmore, B. (1973). *Mon. Not. Roy. Astron. Soc.* **164**, 223.
Ryle, M., and Hewish, A. (1960). *Mon. Not. Roy. Astron. Soc.* **120**, 220.

RYLE, M., and NEVILLE, A. C. (1963). *Mon. Not. Roy. Astron. Soc.* **125,** 39.
RYLE, M., ELSMORE, B., and NEVILLE, A. C. (1965). *Nature (London)* **205,** 1259.
SLEPIAN, D., and POLLAK, H. O. (1961). *Bell Syst. Tech. J.* **40,** 43.
STRATTON, J. A. (1941). "Electromagnetic Theory," McGraw-Hill, New York.
SWARUP, G., and BAGRI, D. S. (1973). *Proc. IEEE* **61,** 1285.
SWARUP, G., THOMPSON, A. R., and BRACEWELL, R. N. (1963). *Astrophys. J.* **138,** 305.
SWENSON, G. W., JR., and MATHUR, N. C. (1968). *Proc. IEEE* **56,** 2114.
THOMPSON, A. R., and BRACEWELL, R. N. (1974). *Astron. J.* **79,** 11.
WADE, C. M. (1970). *Astrophys. J.* **162,** 381.
WEILER, K. W. (1973). *Astron. & Astrophys.* **26,** 403.

Computations in Radio-Frequency Spectroscopy

JOHN A. BALL

CENTER FOR ASTROPHYSICS
HARVARD COLLEGE OBSERVATORY AND
SMITHSONIAN ASTROPHYSICAL OBSERVATORY
CAMBRIDGE, MASSACHUSETTS

I. Introduction to Spectroscopy in Radio Astronomy 177
II. Power Spectra 178
   A. Definitions 178
   B. Estimating Power Spectra in Practice 184
   C. Resolutions and Bandwidths Needed in Practice 197
III. Selected Problems in Calibration and Observing Techniques . . . 200
   A. Notes on Calibrating Radio-Astronomical Data 200
   B. Calibration: $(S-R)/R$ vs. $(S-R)/C$ 202
   C. Switching Techniques and Baseline Curvature 205
   D. Doppler Velocities and the Scale for the Abscissa 207
IV. Selected Problems in Interpretation of Spectra 208
   A. Fitting Gaussians and Baselines 208
   B. Polarization—Stokes Parameters 212
   C. Problems of Multiple Components 214
   References 218

I. Introduction to Spectroscopy in Radio Astronomy

The birth of spectroscopy in radio astronomy is usually dated to 1951 when the 21-cm line of neutral hydrogen was discovered in our galaxy. However, the previous year a swept-frequency spectrometer in Australia was already studying dynamic solar spectra. We now know of over a hundred atoms and molecules with thousands of spectral lines—emission and absorption—from extraterrestrial sources. Pulsars show extraordinarily complex and interesting dynamic spectra. Interest in spectroscopy is intense and a large percentage of all radio astronomers are already preoccupied with acquiring or interpreting spectra. The future promises further outstanding progress if not brilliant discoveries in this field.

The interest in spectroscopy of interstellar and circumstellar atoms and molecules stems from two kinds of problems that can be approached with such measurements. On the one hand, we are interested in the purely astronomical problems of kinematics, dynamics, and distances. The Doppler relation allows us to infer such information from spectra. Also we are interested in the physics and chemistry of atoms and molecules in the interstellar environment. We want to study radiative transfer problems of line formation and questions involving equilibrium or nonequilibrium populations of the levels associated with radio-frequency lines. We would like to know about the formation and destruction and the excitation and ionization of atoms and molecules. The data of spectral-line radio astronomy bear on these problems.

This article includes a discussion of some data-acquisition and data-reduction techniques and some selected problems in data interpretation in spectral-line radio astronomy. However, there is almost no discussion of the actual data nor of particular atoms, molecules, or astronomical sources. There is no discussion of theoretical interpretations of data in terms of astronomical or astrophysical models. There is no advice herein on the actual engineering or construction of hardware. And finally, computer program listings are also omitted even though some such programs relate directly to the problems discussed. All these omitted topics are dealt with in other publications to which references are made at appropriate places herein. By thus limiting the discussion, we aim to provide a deeper analysis within the allotted space.

II. Power Spectra

A. Definitions

One feels intuitively that the power spectrum or power spectral density ought to be something like the Fourier transform of the square of the voltage waveform—at least the units come out right. This Fourier integral usually diverges, however, and there is no simple way to fix it up to be the power spectrum. One also feels intuitively that one ought to be able to measure the power spectrum with a narrow scanning filter followed by a square-law detector. This scheme does work, at least in some approximation depending on the shape of the scanning filter. However, one encounters both mathematical and practical difficulties as one goes to an arbitrarily narrow filter in an attempt to measure the true spectrum.

The alternative approach usually adopted is to define the power spectrum as the Fourier transform of the autocorrelation function of the voltage waveform, and then show that this definition corresponds, at least approximately, to the intuitive approach above. It is also possible to show that the power spectrum is equal to the square of the modulus of the Fourier transform of the voltage waveform, provided this transform exists.

If the voltage waveform is $e(t)$ as a function of time t, then the autocorrelation function $R(\tau)$ is defined as

$$R(\tau) = \lim_{T \to \infty} \frac{1}{2T} \int_{-T}^{T} e(t)\, e(t+\tau)\, dt. \tag{1}$$

The variable τ is called "lag" and has units of time. In practice an infinitely long sample of $e(t)$ is not available so $R(\tau)$ is only known for τ up to some maximum, and only to some approximation that depends on the character of $e(t)$. In radio astronomy one is usually dealing with noiselike voltages characterized by Gaussian statistics. Then $R(\tau)$ calculated for a particular $e(t)$ and finite T can be shown to be a statistically unbiased and convergent estimator of the true $R(\tau)$. The finite maximum value of τ available corresponds exactly to the finite filter width available in the scanning filter scheme above. Thus we have avoided the mathematical difficulty but not the practical one.

The power spectrum is now defined as

$$P(\nu) = \int_{-\infty}^{\infty} R(\tau)\, \exp(-2\pi j \nu \tau)\, d\tau. \tag{2}$$

In this equation ν is frequency and j is $\sqrt{-1}$. Several comments need to be made about this equation. First, the limits on τ will probably be finite as noted above. Also, $R(\tau)$ is both real and symmetrical, so $P(\nu)$ will be also. The transform then becomes just a cosine transform. A 2 before the integral is needed if ν is considered positive only, so $P(\nu)$ is the sum of what would have been equal positive and negative halves. Another 2 is needed if τ also is restricted to be positive.

The usual way of handling the finite range of τ is to set $R(\tau)$ to zero outside the measured range and retain the form of the Fourier integral above. The inverse Fourier transform relation then gives

$$R(\tau) = \int_{-\infty}^{\infty} P(\nu)\, \exp(2\pi j \nu \tau)\, d\nu. \tag{3}$$

It can be shown that a narrow "square" filter with a power transfer function $G(\nu)$ such that

$$G(\nu) = \begin{cases} 1 & \text{for} \quad \nu_0 - \Delta\nu < \nu < \nu_0 + \Delta\nu, \\ 0 & \text{otherwise}, \end{cases} \tag{4}$$

when fed with the $e(t)$ above, produces an output voltage $e_0(t)$ whose power spectrum as defined above is

$$P_0(\nu) = G(\nu)P(\nu) \simeq 2P(\nu_0)\, \Delta\nu, \tag{5}$$

and this is equivalent to showing that our definition of $P(\nu)$ accords with the intuitive scanning-filter ideas above.

Whenever $R(\tau)$ is zero for lags larger than some τ_{max} the Nyquist sampling theorem can be applied to $P(\nu)$. Then values of $P(\nu)$ at discrete points uniformly spaced by $\Delta\nu$ suffice to define $P(\nu)$ exactly, provided $\Delta\nu \leq 1/(2\tau_{max})$. To recover the continuous function $P(\nu)$ from the set of discrete samples of itself $P(\nu_i)$, it is necessary to convolve by sinc, and the convolution becomes a sum, namely

$$P(\nu) = \sum_i P(\nu_i) \text{ sinc} \left(\frac{\nu - \nu_i}{\Delta\nu}\right), \qquad (6)$$

(see, e.g., Bracewell, 1965).

Similarly, $P(\nu)$ is usually restricted in practice by a finite bandpass filter to be (nearly) zero above some frequency ν_{max}. If so, then the Nyquist sampling theoreum applies to $R(\tau)$ also, and it can be sampled at discrete points uniformly spaced by $\Delta\tau \leq 1/(2\nu_{max})$.

Both these sampling relations are valuable in practice because we often wish to use a digital computer or other discrete techniques.

If $R(\tau)$ is measured only up to τ_{max} and if the Fourier transform is taken as if $R(\tau)$ were zero for $\tau > \tau_{max}$, the effect on the spectrum is to convolve it by sinc. This is a special case of a more general relation called the convolution theorem (see, e.g., Bracewell, 1965). A "weighting function" $w(\tau)$ multiplying $R(\tau)$ before the Fourier transform is mathematically equivalent to convolving the spectrum by the Fourier transform of $w(\tau)$. So if

$$P^1(\nu) = \int_{-\infty}^{\infty} w(\tau) R(\tau) \exp(-2\pi j\nu\tau) \, d\tau, \qquad (7)$$

then

$$P^1(\nu) = \int_{-\infty}^{\infty} P(\nu^1) W(\nu - \nu^1) \, d\nu^1, \qquad (8)$$

where

$$W(\nu) = \int_{-\infty}^{\infty} w(\tau) \exp(-2\pi j\nu\tau) \, d\tau, \qquad (9)$$

and where $P(\nu)$ is from Eq. (2) above. The simplest weighting function, called uniform weighting, is unity for τ up to τ_{max} and zero beyond, and its Fourier transform is sinc.

If the spectrum is measured with a scanning filter (described in more detail in Section II, B below) then the weighting function is just the Fourier transform of the filter shape. However, in this case there is usually no need to work in lag space at all (but see Section IV, C).

The point is that any spectrometer is characterized by some weighting function or other, or its corresponding convolving function—the resolution.* The true spectrum is never available in practice and sometimes not even in principle. One has a choice among various weighting functions, but one cannot choose no weighting function at all.

Table I contains information about six popular weighting functions, and Fig. 1 shows the resolution functions for two of these. Instead of one best weighting function for all applications, we have a variety to choose from, depending on the particular purpose in mind.

For a given τ_{max}, uniform weighting can be shown to produce the best approximation to the true spectrum in a least-squares sense and to produce the highest signal to noise ratio for narrow (unresolved) features. Furthermore, uniform weighting can reproduce the true spectrum if it is band-limited at less than the τ_{max} of the spectrometer, i.e., if the spectrum is resolved.† However, the very high sidelobes produced by uniform weighting may be mistaken for additional features especially if the spectrum is narrow but complex. Most users prefer a weighting function that produces low sidelobes to avoid this and related confusion problems. An exactly analogous problem occurs in two dimensions in mapping on the sky.

Regardless of how the spectrum is obtained, the resolution function can be changed by an additional convolution perhaps performed during the data-reduction process. This additional convolution can either widen (called "smoothing") or narrow (called "restoration") the resolution and

---

* We shall use the term "bandwidth" to refer to the whole frequency range that the spectrometer analyzes and "resolution" or "resolution function" to refer to the scanning or convolving function in frequency space. The term "spectral window" has, unfortunately, been used for both bandwidth and resolution, is therefore ambiguous, and will not be used herein.

Other ambiguous terms that we shall avoid are "higher" and "lower" as applied to resolutions; we shall use "wider" and "narrower" instead.

† The words "resolved" and "unresolved" are used very loosely in radio astronomy. If a spectrum or spectral feature is referred to as "unresolved" it might mean either that (a) additional detail in the spectrum or map would be obtained by using narrower spectral resolution, (b) a feature width less than (or much less than) the instrumental resolution is permitted by the data, or (c) a feature width less than (or much less than) the instrumental resolution is demanded by the data, etc. Furthermore, the feature width is itself difficult of definition unless the spectrum is particularly simple.

We suggest that the term "resolved" be applied to a spectrum whose Fourier transform is band limited with a maximum lag less than the limit of flat response characterizing the spectrometer; so, aside from noise considerations, a spectrometer produces an accurate replica of a resolved spectrum but not of an unresolved one. For an autocorrelation spectrometer with uniform weighting, the limit of flat response is just τ_{max}. With this definition, (a) above is true, but (b) and (c) are still ambiguous.

TABLE I
SIX POPULAR WEIGHTING AND RESOLUTION FUNCTIONS

Name	Weighting function[a]	Resolution function	Resolution width[b]	Peak sidelobe level (%)
Uniform[c]	1	$2\tau_{max} \text{ sinc } (2\nu\tau_{max})$	0.60	-22
Cosine	$\cos [\pi\tau/(2\tau_{max})]$	$\tau_{max} [\text{sinc } (2\nu\tau_{max} + \frac{1}{2})$ $+ \text{sinc } (2\nu\tau_{max} - \frac{1}{2})]$	0.82	-7
Triangle (Bartlett)	$1 - |\tau|/\tau_{max}$	$\tau_{max} \text{ sinc}^2 (\nu\tau_{max})$	0.89	$+4.7$
Hanning (Cosine squared)	$\cos^2 [\pi\tau/(2\tau_{max})]$ $= 0.5 + 0.5 \cos (\pi\tau/\tau_{max})$	$\tau_{max}[\text{sinc } (2\nu\tau_{max})$ $+ 0.5 \text{ sinc } (2\nu\tau_{max} + 1)$ $+ 0.5 \text{ sinc } (2\nu\tau_{max} - 1)]$	1.00	-2.7
Hamming	$0.54 + 0.46 \cos (\pi\tau/\tau_{max})$	$\tau_{max}[1.08 \text{ sinc } (2\nu\tau_{max})$ $+ 0.46 \text{ sinc } (2\nu\tau_{max} + 1)$ $+ 0.46 \text{ sinc } (2\nu\tau_{max} - 1)]$	0.91	± 0.74
Blackman[d]	$0.42 + 0.5 \cos (\pi\tau/\tau_{max})$ $+ 0.08 \cos (2\pi\tau/\tau_{max})$	$\tau_{max}[0.84(\text{sinc } 2\nu\tau_{max})$ $+ 0.5 \text{ sinc } (2\nu\tau_{max} + 1)$ $+ 0.5 \text{ sinc } (2\nu\tau_{max} - 1)$ $+ 0.04 \text{ sinc } (2\nu\tau_{max} + 2)$ $+ 0.04 \text{ sinc } (2\nu\tau_{max} - 2)]$	1.15	± 0.12

[a] All weighting functions are zero for $\tau > \tau_{max}$.
[b] The full width to half maximum in units of $1/\tau_{max}$.
[c] Uniform weighting corresponds to the "principle solution" of Bracewell (1965).
[d] Blackman and Tukey (1958).

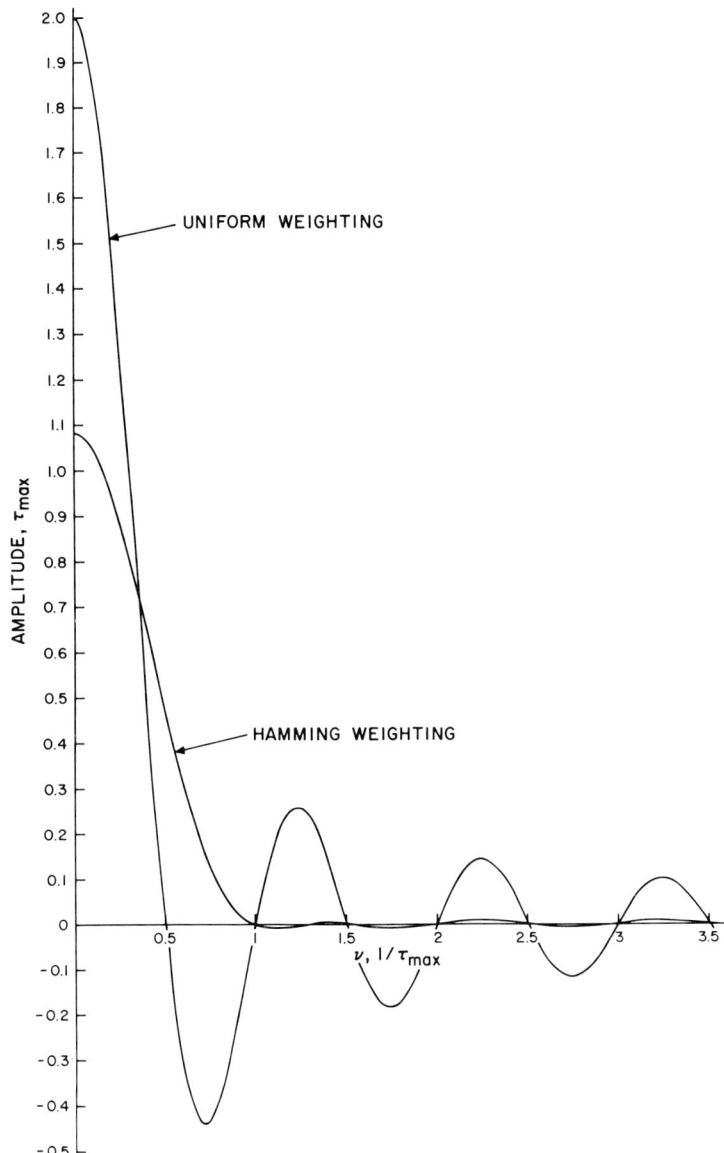

Fig. 1. Two popular resolution functions. The abscissa is in units of $1/\tau_{max}$. The Nyquist interval is 0.5 in these units. The weightings are explained in Table I.

also affect the sidelobe level. However, the range of control available through this technique is limited because there is no way to obtain information not already contained in the data.

By the convolution theorem, this additional convolution is equivalent to multiplication by an additional weighting function in lag space. So one can change the weighting function arbitrarily except at lags where the original weighting function is zero. Note that the resolution functions in Table I are the spectrometer's response to a narrow spectral line, but also the functions to be convolved with spectra from uniform weighting to produce the corresponding resolution shapes.

If the resolution is narrow compared with the narrowest feature or detail on the spectrum, then the effect of the resolution is usually negligible. But if the resolution is comparable to or wider than the feature width, then its effect on the spectrum can be calculated in some simple cases. For example, if both the resolution and the feature are approximated as Gaussian in shape, then the width of the response on the observed spectrum can be shown to be

$$W_R \simeq (W_F^2 + \beta^2)^{1/2}, \tag{10}$$

where W_F is the width of the feature and β is the spectrometer resolution.

Figure 2, based on this equation, shows the effect on the spectrum of using a resolution comparable to the feature width. A resolution wider than about a quarter of the feature width is incompatible with precise measurements unless one somehow compensates for the smearing. For isolated Gaussian features with little noise, this compensation can sometimes be done just using these curves or Eq. (10) above; but for complex or overlapping features, a compensation procedure is more difficult or impossible. It is preferable to use a sufficiently narrow resolution to avoid the effects of smearing. This problem is discussed further in Section II, B, 5.

B. Estimating Power Spectra in Practice

1. Scanning Filters

Both the intuitive approach to defining power spectra and the first measurements of spectra in radio astronomy made use of scanning filter spectrometers. One selects a filter whose width (the resolution) is narrow enough not to smear the spectral details, connects the output of the filter to a square-law (power) detector and an integrator, and scans the frequency of the filter across the bandwidth of interest. The output of such a system is a voltage proportional to the weighted average power spectrum over the width of the filter. In practice the filter is usually fixed, but the local oscillator frequency is varied in a superheterodyne configuration to achieve the effect of a variable-frequency filter. Figure 3 is a block diagram of such a receiver.

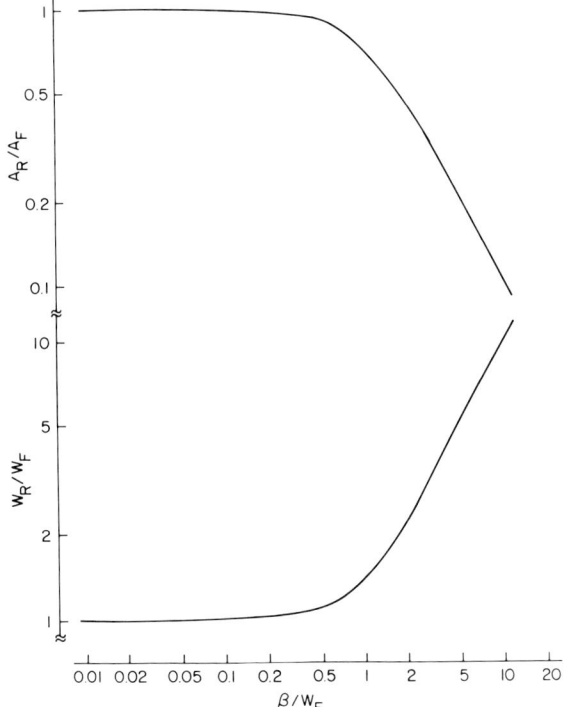

Fig. 2. Degradation of a Gaussian feature in amplitude (above) and width (below) due to finite spectrometer resolution. A_R and W_R are the amplitude and width of the spectrometer response; A_F and W_F are the true feature amplitude and width for an arbitrarily narrow β—the spectrometer resolution.

The spectrum as measured by such a system is the convolution of the true spectrum on the sky by the resolution, namely,

$$P^1(\nu) = \int_{-\infty}^{\infty} H(\nu - \nu^1) P(\nu^1) \, d\nu^1, \tag{11}$$

where $H(\nu)$ is the power transfer function of the resolution filter. So $P^1(\nu)$ is band-limited in lag space if (a) the true spectrum on the sky, $P(\nu)$, is so limited, or if (b) the resolution $H(\nu)$ is such that its Fourier transform is, at least approximately, band-limited. In some cases, for example if one has relatively wide "square" filters, neither of these conditions even approximately holds. The simplest usable filter shape (single-tuned RLC) has a Fourier transform that decreases exponentially but never quite reaches zero. However, if either of these conditions does hold, then one can step rather than scan the filter across the bandwidth in accordance with the Nyquist sampling theorem.

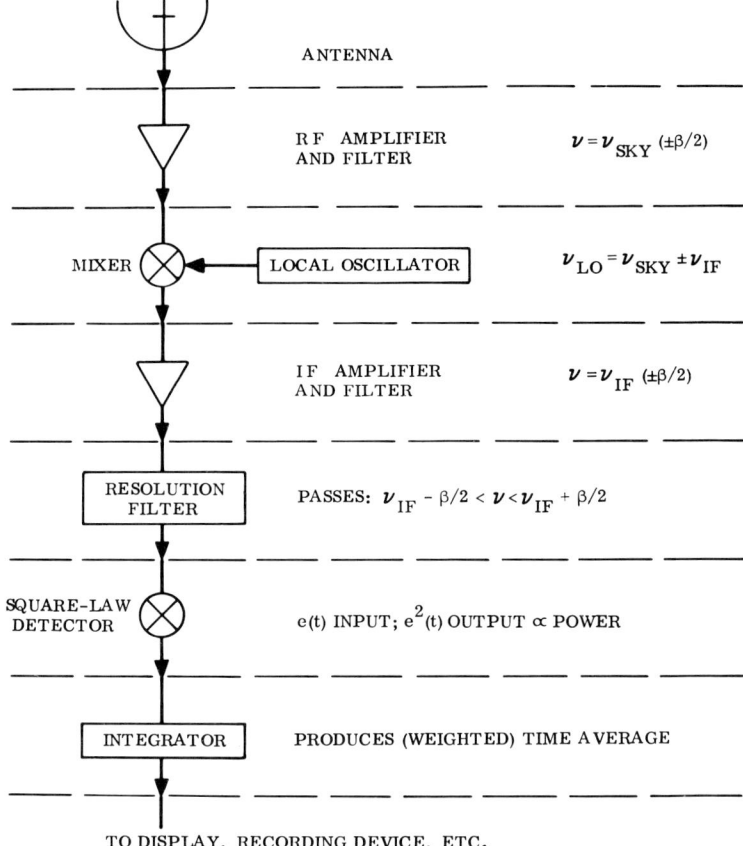

Fig. 3. Block diagram of a scanning filter receiver.

Most signals to be measured in radio astronomy are noiselike and characterized by Gaussian statistics. If so, then the envelope of a narrow-band sample of such noise has a flat spectrum and it can be shown that the fractional rms fluctuation level of the detected power in such a signal is given by $1/\sqrt{(\beta t)}$, where β is the noise width of the resolution filter, and t is the integration or averaging time (see Rice, 1954, Section 3.9).

Many different forms of this equation occur in radio astronomy. We need only note here that if noise is a problem, as it usually is, then the scanning filter spectrometer is inefficient in that it measures the spectrum at only one point at a time. In effect, all the information coming through the bandpass but outside the resolution is ignored and wasted. In the noise equation above, t can be no more than $1/N$th of the total observing session, where N is the number of steps across the spectrum.

2. Multichannel Filter Systems

We are led, then, to consider the multichannel filter spectrometer. The additional complexity is the price one pays for higher observing efficiency or a lower rms noise for a given observing session.

In designing a comb of filters, we want to use the Nyquist sampling theorem to determine the filter spacing, and the conditions necessary for the sampling theorem to apply are just the same as we considered in Section II, B, 1 above. If a wider spacing is used then some information will be lost. However, it may be recoverable through additional measurements made with the whole comb of filters stepped by small amounts in frequency. The observing efficiency is clearly the highest when one has enough filters to satisfy the Nyquist criterion, and also the spectrum from a multichannel filter system will then be equivalent to the corresponding spectrum from a scanning filter system [cf. Eq. (6) above].

Figure 4 is a block diagram of a multichannel filter receiver. Each of the filters in the filter bank must have its own square-law (power) detector and integrator (averager). An alternative to the scheme in Fig. 4 is a set of identical filters each with its own mixer and local oscillator,

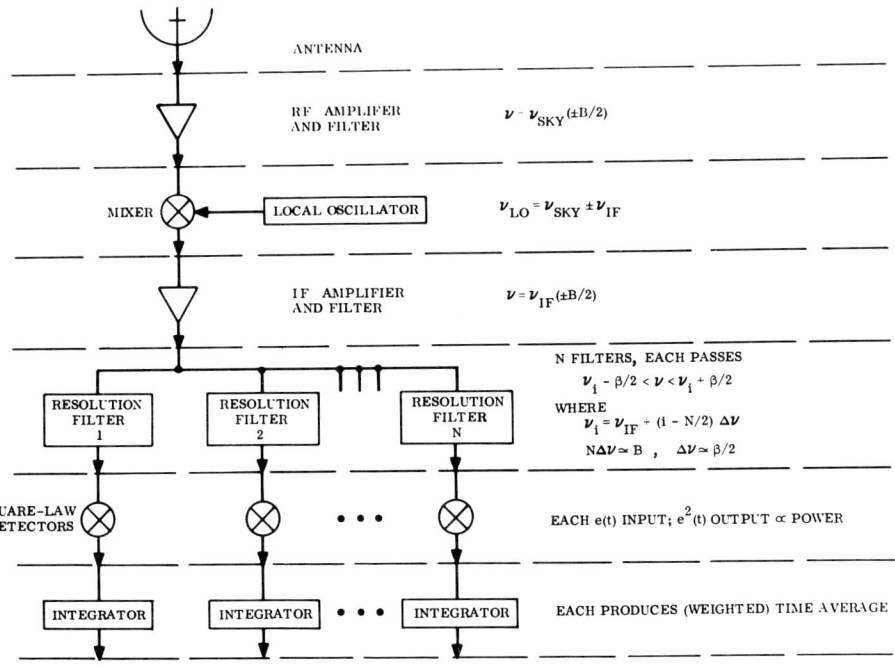

FIG. 4. Block diagram of a multichannel filter receiver.

with the local oscillator frequencies all different so as to produce the effect of a comb of filters at the signal frequency.

Multichannel filter spectrometers have been built with as many as 256 filters. Because the center frequency of the filter bank can be chosen freely in a superheterodyne system, an extremely wide range of filter resolutions and filter shapes is available in practice. For these and other reasons, multichannel filter receivers are probably more widely used in radio astronomy than any other type of spectrometer.

Neither Fig. 3 nor 4 includes an explicit scheme for switching. Instead, switching schemes applicable to all types of spectrometers are discussed separately in Section III, C.

3. Acoustooptic (Bragg-Cell) Spectrometers

The principle of a Bragg-cell spectrometer is certainly not new (see, e.g., Gordon, 1966), however, the development of a practical device for use in radio astronomy has come only recently (Cole, 1968a, 1973a,b; Hecht, 1973; Cole and Ables, 1975).

Figure 5 is a schematic diagram of such a device. The laser light is fixed in wavelength, amplitude, and direction. The signal to be analyzed is fed to a piezoelectric transducer on the Bragg cell and so converted to traveling acoustical waves in the cell. These are compression waves and the material of the cell is such that the index of refraction is affected by compression. Thus, the emerging light is phase modulated across the wavefront and is capable of interfering with itself.

The result is that some of the laser light is diffracted through an angle that depends on the acoustical wavelength in the Bragg cell, i.e., the

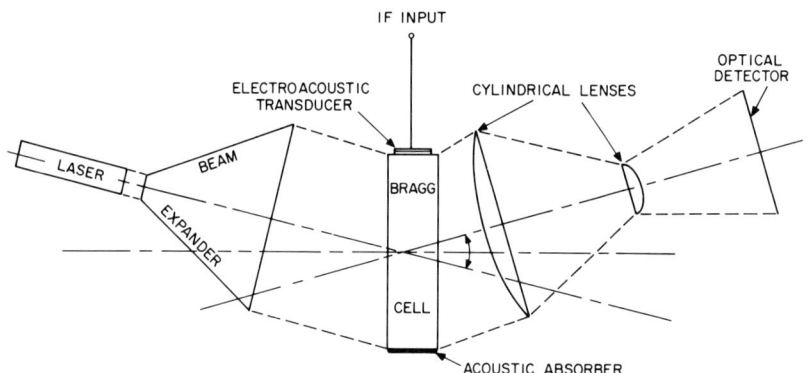

FIG. 5. Schematic diagram of a Bragg-cell spectrometer.

wavelength of the signal to be analyzed. The diffraction angle is approximately equal to λ/Λ, where λ is the wavelength of the laser light in the material of the Bragg cell and Λ is the acoustical wavelength. This relation can be derived either from the classical wave diffraction approach or by considering photon–phonon collisions. The collision approach gives only the first-order spectrum and so indicates that only this order will be appreciable if the cell is large in wavelengths for both the optical and acoustical waves.

It can be shown that the brightness of the diffracted light will be approximately proportional to the square of the amplitude of the acoustical wave provided only a small percentage of the laser light is diffracted. It can also be shown that there is almost no interaction between two signals simultaneously present provided they are not closer together than the resolution, farther apart than an octave, nor so strong that a large percentage of the laser light is diffracted. Within these limitations, then, the Bragg-cell device converts a radio-frequency spectrum into an intensity modulated strip of light which can be recorded with any of a variety of types of optical sensors.

The properties needed in this optical sensor depend on the application. For recording dynamic spectra such as solar bursts or pulsars, a moving-film camera is appropriate. If further processing in a computer is needed, as is usually the case, then one can use a TV vidicon tube or an array of photodiodes to convert the spectrum to a voltage which can then be digitized for a computer.

The resolution of a Bragg-cell spectrometer is determined largely by the number of acoustical wavelengths through the region in which the acoustical and optical waves interact. Thus narrowing the resolution involves lengthening the Bragg cell or making it of a material in which the speed of sound is slower. The attenuation of the acoustical wave through the cell can also affect the resolution by shortening the effective interaction length. The resolution can also be degraded, of course, by imperfections in the laser, the optics, or the optical sensor at the output.

Bragg-cell spectrometers that analyze a 100 MHz bandwidth with about 125 kHz resolution (Cole, 1973a,b) and a 1.5 MHz bandwidth with about 20 kHz resolution (Cole and Ables, 1975) have already been used in radio astronomy. Both wider bandwidths and narrower resolutions are possible and will probably be available soon.

Bragg-cell spectrometers show very considerable promise especially for use with millimeter-wavelength front ends. Spectra in this wavelength range are proving to be very complex, interesting, and of course, quite wide in frequency. Autocorrelation spectrometers for this bandwidth range are still prohibitively complex and expensive, and filter receivers only slightly less so. Thus the Bragg-cell spectrometer promises to dominate this and perhaps other areas in spectral-line radio astronomy in the near future.

4. Autocorrelation Spectrometers

Autocorrelation spectrometers make direct use of the definition of a power spectrum given in Section II, A. The autocorrelation function of the voltage waveform can be computed either by hardware (an "autocorrelator"), or software (a computer program in a general purpose computer); and the correlation can be done on a full multibit or analog representation of the voltage, or on a representation containing very few bits—perhaps only one. The Fourier transform is then performed, usually in a general purpose computer, to give the spectrum.

The most popular autocorrelation systems in radio astronomy today utilize one-bit or hard-clipped autocorrelators (Weinreb, 1963). It can be shown that the Fourier transform of the sine of the autocorrelation function of the clipped voltage waveform is an unbiased estimate of the power spectrum of the unclipped signal provided it is a Gaussian random variable (Van Vleck, 1943; Clark, 1970). So we have

$$R(\tau) \simeq \sin[(\pi/2) R_c(\tau)], \qquad (12)$$

where $R(\tau)$ is the ordinary autocorrelation function and $R_c(\tau)$ is the autocorrelation function of the clipped voltage, namely,

$$R_c(\tau) = \lim_{T \to \infty} \frac{1}{2T} \int_{-T}^{T} y(t) y(t+\tau) \, dt, \qquad (13)$$

where

$$y(t) = \begin{cases} +1 & \text{if } e(t) > 0, \\ -1 & \text{if } e(t) < 0. \end{cases} \qquad (14)$$

A one-bit correlator is simpler and computationally faster because the one-bit multiplier is just a negated exclusive-OR gate (NEXOR) and the adder is just a counter. The sine clipping correction requires only a small amount of additional computation time.

The computational speed and simplicity associated with the one-bit representation are advantages also if the autocorrelation is performed in software. However, the software approach is not very attractive in any case because it is limited to slow data rates, i.e., narrow bandwidths. It may offer a quick solution if a computer is available and free of other tasks.

A one-bit autocorrelation system as used in radio astronomy is shown schematically in Fig. 6. The signal is converted to "video," i.e., to a fre-

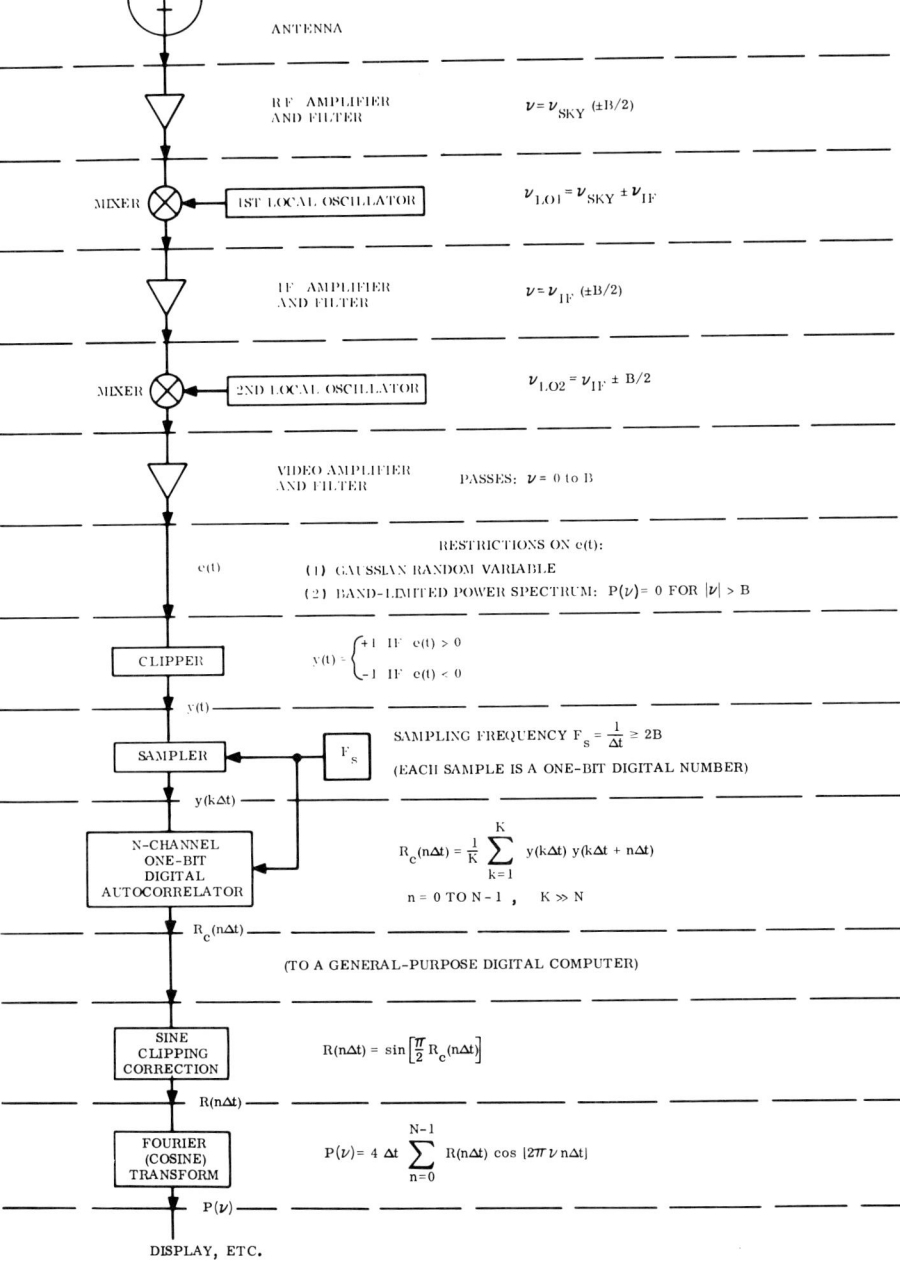

FIG. 6. Block diagram of a one-bit digital autocorrelation receiver [based on Fig. 7 in Weinreb (1963)].

quency range extending from zero frequency upward. The clock rate or sampling frequency F_s is then chosen by the Nyquist criterion to be at least twice the video bandwidth B, i.e., twice the bandwidth to be analyzed. The spectral resolution β (for Hanning weighting) is then equal to this sampling frequency divided by N, the number of channels (i.e., the number of lags computed) in the correlator. In conventional correlation systems the number of channels is fixed by the hardware, but the bandwidth and sampling frequency are selectable by the user.

The cosine transform needed to get to the power spectrum can be performed in a general purpose digital computer using some version of the fast Fourier transform (FFT) algorithm. This algorithm in its common form is, however, a full complex Fourier transform and so is wasteful of both storage space and computational speed. A version of the FFT algorithm is available to do just the cosine transform (e.g., Brenner, 1973) with little wasted space or speed; however, the program itself is rather longer than an equivalent complex form.

Fortunately the time necessary to perform the transform is rarely a concern because the transform needs to be done only once (or twice) per spectrum—usually only every few minutes or longer. It is even possible, though not necessarily desirable, to record the autocorrelation numbers and do the transform later.

Conventional one-bit digital correlators as used in radio astronomy are inefficient in the sense that the digital logic of which they are built is being used at its maximum speed only when operating at the widest bandwidth. At slower sampling frequencies, in effect, computational power is being wasted. If temporary storage is available for blocks of raw data bits, then a correlator system can be built that recycles the data through the correlator several times over at slow data rates (i.e., narrow bandwidths) to multiply the number of hardware channels by a factor equal to the number of times recycling takes place. Such a system is described by Ball (1973).

The price one pays for computational simplicity and speed in working with a one-bit representation of the signal is added noise. The noise performance of a one-bit correlator is poorer than that of a multibit correlator or an equivalent filter receiver because the process of representing the signal only by its sign introduces quantization noise. A good discussion of this point is given by Burns and Yao (1969). The noise is increased by a factor of about $\pi/2$, however, this factor depends on the shape of the bandpass preceding the clipper and on the exact sampling frequency. Oversampling, that is, sampling at a rate more than twice the bandpass, reduces the excess noise somewhat.

Multibit correlators with improved noise performance have been proposed, studied, and built (Cole, 1968b; Cooper, 1970; Hagen and Farley, 1973; Bowers et al., 1973; Bowers and Klinger, 1974). However, such correlators are more complex than one-bit designs, and it is widely believed that increasing complexity should be such as to provide more channels and/or wider bandwidths rather than a small improvement in noise performance.

5. Maximum Entropy Method and Related Techniques

The maximum entropy method (MEM) is the best-known example of a class of techniques that attempt to circumvent, in part, the difficulties associated with the lack of knowledge of the autocorrelation function for lags larger than τ_{max}. MEM is an alternative to the conventional scheme which takes the Fourier transform of the (weighted) autocorrelation function as an estimate of the power spectrum. The computations necessary to get from the autocorrelation numbers to the MEM spectral estimate are only slightly more complex than the conventional scheme and so are not a serious drawback. Under certain circumstances MEM promises "super-resolution," i.e., narrower resolution than can be obtained with the conventional scheme on the same autocorrelation numbers.

The inverse Fourier transform of the MEM spectral estimate is equal to the original autocorrelation numbers for lags up to the maximum lag measured. This is exactly equivalent to the conventional scheme with uniform weighting. However, the inverse Fourier transform of the MEM spectrum is *not* restricted to be zero for larger lags as is the case in the conventional scheme. So MEM has somehow "guessed" some of the unmeasured autocorrelation numbers for larger lags. The differences among MEM and several related techniques are just the criteria for this guessing.

To write the MEM equations, we need to convert our previous Fourier transforms to their discrete form. In place of Eq. (2), we write

$$P_i = \frac{1}{(2M)^{1/2}} \sum_{k=-N}^{N} R_k \exp(-\pi jki/M). \tag{15}$$

In this equation i runs from $-M$ to M and j is $\sqrt{-1}$. In place of Eq. (3), we write

$$R_k = \frac{1}{(2M)^{1/2}} \sum_{i=-M}^{M} P_i \exp(\pi jki/M). \tag{16}$$

Finite limits are put on k ($-N$ to N) because it is just the problem of the unknown R_k for $|k| > N$ that MEM deals with.

The MEM spectral estimate S_i is also constrained to satisfy

$$R_k = \frac{1}{(2M)^{1/2}} \sum_{i=-M}^{M} S_i \exp(\pi j k i/M). \tag{17}$$

However, $|k| > N$ in Eq. (16) gives zero, but not so for Eq. (17) provided $M > N$ as is necessary for MEM. S_i is not band-limited as is P_i.

If S_i is thought of as being made from a flat spectrum passed through a filter, then the entropy gain of the filter is proportional to

$$I = \sum_i \log S_i. \tag{18}$$

We want to maximize I subject to the constraints in Eq. (17) above, using the technique of Lagrangian multipliers. Define

$$Q = \sum_i \log S_i + \sum_k \mu_k \left[R_k - \frac{1}{(2M)^{1/2}} \sum_i S_i \exp(\pi j k i/M) \right], \tag{19}$$

where μ_k are the Lagrangian multipliers to be determined. We differentiate Q with respect to each S_i and set the result to zero for a maximum. This gives

$$\frac{1}{S_i} = \frac{1}{(2M)^{1/2}} \sum_{k=-N}^{N} \mu_k \exp(\pi j k i/M). \tag{20}$$

So although S_i is not band-limited, its reciprocal is. The problem now is to find the μ_k (or the S_i) given the R_k (or the P_i). There are as many μ_k as R_k, so the problem may be determinate even though $M > N$.

A solution to this particular MEM problem is available in closed form. It involves rewriting Eq. (20) as

$$\frac{1}{S_i} = \frac{1}{2M} \left| \sum_{k=0}^{N} \Gamma_k \exp(\pi j k i/M) \right|^2. \tag{21}$$

This is no additional restriction because S_i must be real, symmetrical, and positive anyway. Then it can be shown, using some rather messy expansions, that the Γ_k satisfy

$$\sum_{k=0}^{N} \Gamma_k R_{k-l} = \delta_l, \tag{22}$$

where δ_l is 1 if l is zero, and zero otherwise. In this equation l is construed as positive only, and the sum runs over positive k only as indicated. Again this is no additional restriction because R_k is symmetrical anyway.

The R_{k-l} in Eq. (22) form a Toeplitz matrix and this matrix equation can be inverted using an algorithm for Toeplitz matrices suggested by Levinson (1947) to give Γ_k to put into Eq. (21).

Equation (22) above looks very much like a convolution, and it might be possible to use the convolution theorem to solve for the Γ_k or their Fourier transform. However, no one seems to have worked out this alternative formulation.

Lacoss (1971) applied MEM to various artificial test functions and to data on seismic Earth motions with promising results. MEM was applied to spectra in radio astronomy by Ables (1974) and by Ball (1972).

Under the following circumstances MEM degenerates almost exactly into the conventional scheme with uniform weighting (i.e., almost no additional autocorrelation numbers are guessed): (a) a complex spectrum (many overlapping features) regardless of the signal-to-noise ratio (SNR), or (b) a noisy spectrum (SNR~3 or less) regardless of spectral complexity. For a resolved spectrum (regardless of complexity or SNR), MEM offers no advantages over the conventional method; it usually differs only in the noise, and only very slightly.

So there is a very narrow range of circumstances in spectral-line radio astronomy for which MEM offers any advantages, namely, a simple spectrum (one or two features, say) which is only partially resolved but which is seen with a good SNR (say 5 or more—the more the better). For this case MEM can do much better than the conventional scheme. MEM sometimes gives a much better estimate of the width and amplitude of an isolated unresolved feature and·MEM sometimes separates blended features unresolved by the conventional scheme. In this case MEM will also reduce or eliminate the spectral sidelobes associated with uniform weighting.

The circumstances under which MEM offers some advantages are just those in which one might be tempted to correct the observed feature widths and amplitudes for instrumental resolution. MEM makes such corrections automatically.

It has also been argued (for example, by Ables, 1974) that MEM is preferable *in principle* to the traditional scheme. MEM produces a spectrum that is the convolution of the true spectrum by a convolving function whose width and shape are not known and in fact vary from one spectrum to another and even from one point to another within the same spectrum. The traditional scheme, as discussed in Section II, A, produces a spectrum that is the convolution of the true spectrum by a specific known convolving function. In case the spectrum is unresolved, the traditional spectrum *cannot* be the true spectrum; the MEM spectrum *might* be. The traditional spectrum is, however, obtained from the true spectrum by a definite, easy to visualize procedure; the MEM spectrum by a definite but abstruse procedure. Some radio astronomers prefer the former *in principle*.

And MEM can be fooled. A procedure for doing this deliberately is just to give MEM any resolved spectrum, i.e., one for which R_k *is* zero for $|k| > N$, then any nonzero R_k guessed by MEM will be wrong. It is possible to construct artificial test functions for which the MEM spectral estimate is disastrously wrong; these functional forms are perhaps unlikely but not impossible in practice.

These comments are not so much a criticism of MEM or of the derivation of the MEM formulas as of the somewhat exaggerated claims made for MEM.

The advantages offered by MEM turn out to be more important in mapping (either with a single antenna or with an interferometer) because the circumstances under which MEM excels occur more frequently and more inevitably. Rogers (1974) studied two-dimensional MEM in some detail. From his work it seems that if one must guess autocorrelation numbers that were not measured, then at least MEM is a good—maybe the best—way of guessing.

6. Direct Fourier Transforms

In Section II, A above, we noted that the power spectrum could also be obtained from the square of the modulus of the Fourier transform of the voltage waveform,

$$P(\nu) \simeq \frac{1}{T} \left| \int_0^T e(t) \exp(2\pi j\nu t) \, dt \right|^2. \tag{23}$$

This direct Fourier transform can be performed by a computer program in a general purpose computer ("software FFT") or by a hardware fast Fourier transform processor ("FFT machine," Yen, 1974). The software approach is not very attractive because it is limited to very slow data rates, i.e., very narrow bandwidths, although it may offer a quick solution if a computer is available and free of other tasks. And the FFT machine tends to be rather complicated to build especially for fast data rates. If only a spectrum is wanted, the FFT machine probably offers no advantages over other schemes. However, if voltage cross-correlations—multiplications in transform space—are also needed, as in an interferometer, then the FFT machine may offer some advantages.

C. Resolutions and Bandwidths Needed in Practice

In this section we give a very brief discussion of known properties of some radio sources on the sky, feature widths and spectral coverage, to establish approximate numbers for the resolutions and bandwidths needed in a spectrometer for radio astronomy. We use a result discussed in Section II, A, namely, that a resolution wider than about a quarter of the feature width is undesirable.

It is usually convenient to think of resolutions and bandwidths in velocity units (km/sec) because then one can often characterize a given source without reference to the molecule or transition observed. For example, the velocity resolution and bandwidth needed to study recombination lines toward a given source are nearly independent of wavelength or principle quantum number n. However, most spectrometers actually operate in frequency units and the observer needs to select bandwidths and resolutions from among those available in practical observing situations.

Figure 7 is a summary of bandwidths and resolutions needed to observe certain well-known sources. The approximate nature of this figure should be emphasized. New discoveries of other emission or absorption features may be expected and new detail may be discovered in known features as more precise measurements are made in the future. Both wider bandwidths and narrower resolutions may be needed not only for new sources yet undiscovered, but even for the familiar sources in Fig. 7.

The numbered lines in Fig. 7 correspond to the number of filters or autocorrelation channels necessary for a spectrometer to make the corresponding measurements. These numbers are based on an approximate Nyquist criterion of two data per resolution width.

The dynamic spectra of pulsars and solar bursts prove to be very interesting. The spectrometer for such applications records intensity as a function of frequency and time. A useful display is moving film on which darkening represents intensity, and frequency or wavelength is plotted across the film, and time along the length of the film. A Bragg-cell spectrometer produces such a display directly, and the first use of the Bragg-cell spectrometer in radio astronomy was to study dynamic solar spectra (Cole, 1973a). A multichannel or scanning filter spectrometer can also be used if its output is converted to modulated light using, for example, a cathode ray tube.

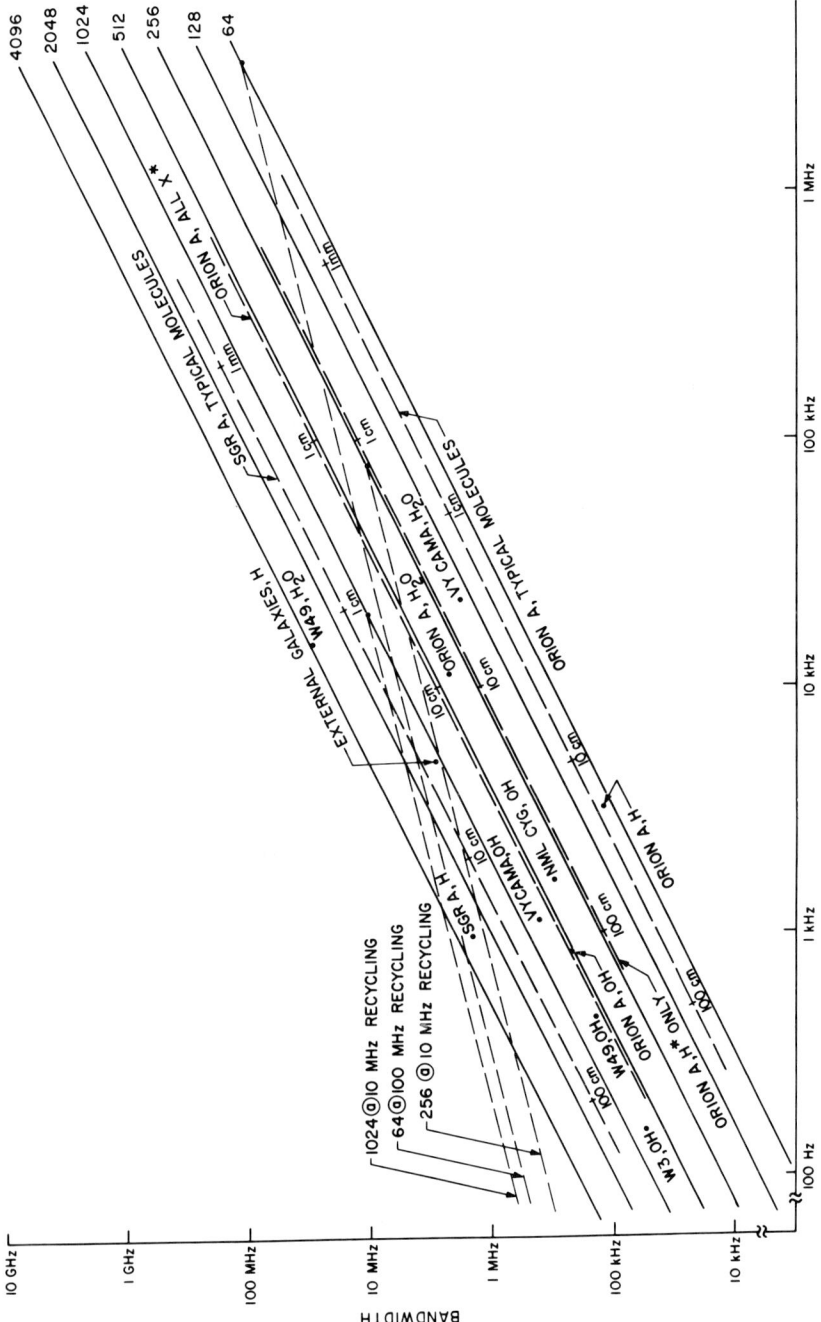

FIG. 7. Approximate bandwidths and resolutions needed in practice. The numbered lines correspond to a spectrometer, with the indicated number of channels, following an approximate Nyquist criterion of two data per resolution width. The resolutions for various sources are about a quarter of the narrowest (known) feature width (cf. Fig. 2).

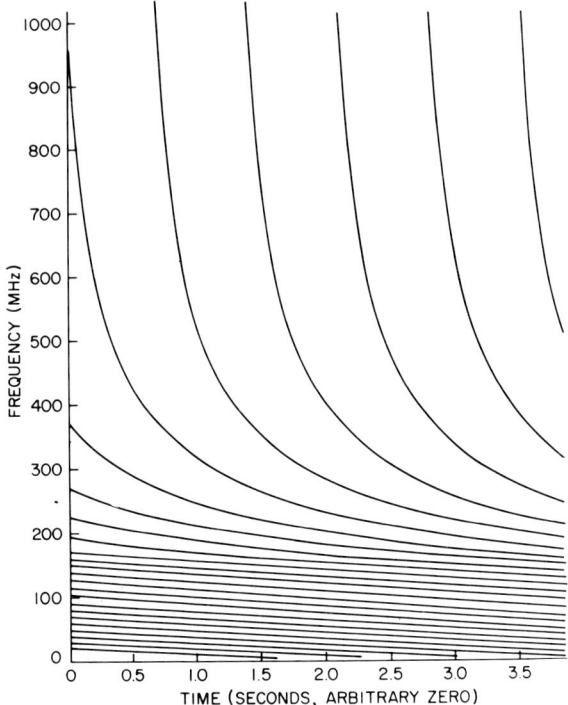

FIG. 8. Calculated dynamic spectrum of a typical pulsar, CP0329.

On a moving-film photographic display pulsars appear as corrugations tilted with respect to the direction of motion of the film. The angle of tilt depends on the dispersion. The human eye is quite skillful in picking such patterns out of noise. Known pulsar periods range from 0.3 to 4 sec and dispersions from 1 to 25×10^{16} Hz. CP0329 is a typical pulsar with a period of 0.7145 sec and a dispersion of 11.1×10^{16} Hz. Figure 8 shows a calculated display of this type for CP0329. The abscissa is time and corresponds to the motion of the film. The ordinate is frequency over the range that CP0329 is strongest. One should imagine the lines in this figure smeared into broad erratic bands to approximate real observations.

Figure 9 shows the resolution required for a given time smearing $\Delta \tau$ for this same pulsar CP0329. This figure is calculated on the assumption that the local oscillator frequency is fixed rather than scanning to follow the dispersion.

For a typical pulsar a bandwidth of 400–700 MHz would be desirable, but the resolution would need to be very narrow indeed at low frequencies in order to avoid time smearing. No such spectrometer is available yet; observations of pulsars are usually done with much narrower bandwidths.

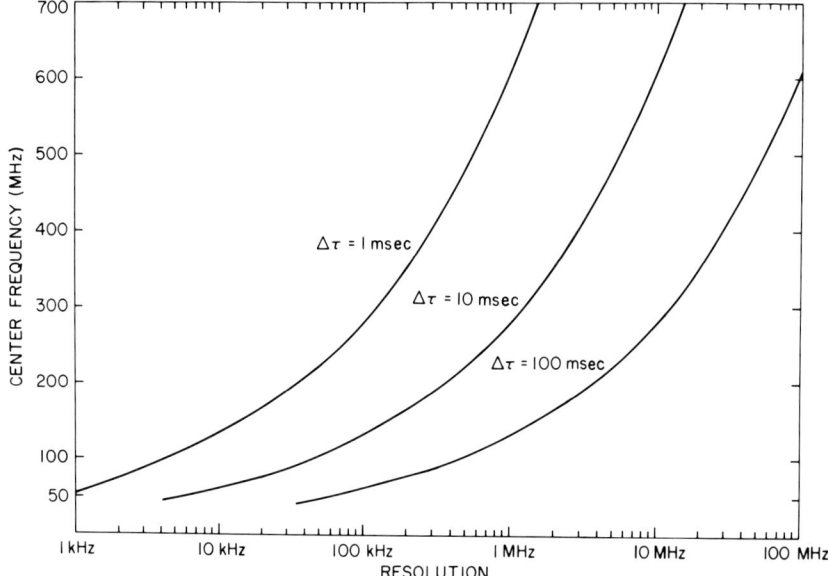

FIG. 9. Spectral resolution required for a given time smearing on CP0329.

III. Selected Problems in Calibration and Observing Techniques

A. Notes on Calibrating Radio-Astronomical Data

One would think that the problems of calibrating in radio astronomy would be so well known that any additional discussion now would be a pointless duplication. However, it is not well known that one can acquire and calibrate radio-astronomical data without knowing the value of the calibration noise tube in units of antenna temperature, or indeed without having a calibration noise tube at all. One cannot, in this case, determine an aperture or a beam efficiency, but, as outlined below, astronomers (as distinguished from engineers), do not need to know these quantities.

The true brightness temperature is equal to the apparent brightness temperature for any source whose angular structure is not smeared by the antenna beam, i.e., for a source that is completely resolved. The true flux density is equal to the (peak) apparent flux density for any source that is completely unresolved, i.e., for a source that is small compared with the antenna beam. The goal of the calibration procedure is to place an apparent brightness temperature scale and an apparent flux density scale on the ordinate of the scans or spectra.

If an extended radio source of known brightness temperature is available for calibration, then the brightness temperature scale (i.e., the system temperature and/or the calibration noise tube temperature in units of apparent brightness temperature) can be measured directly. More typically, however, the calibration available is a small-diameter radio source of known flux density from which the apparent flux-density scale (i.e., the system temperature and/or the calibration noise tube temperature in units of apparent flux density) can be measured directly.

Thus the typical calibration problem involves only the additional step of converting the apparent flux-density scale to an apparent brightness temperature scale. This scale conversion is done by multiplying by a constant which depends on the wavelength and Ω_B, the measured beam solid angle of the telescope. In particular

$$T_{B,\,Ap}(\Omega) = \frac{2k}{\lambda^2} \frac{S_{\nu,\,Ap}(\Omega)}{\Omega_B}. \tag{24}$$

where

$$\Omega_B \simeq (1.133)\theta_\alpha \theta_\delta \tag{25}$$

and where θ_α and θ_δ are the measured halfwidths of the antenna beam in right ascension (α) and declination (δ). The constant 1.133 comes from assuming that the beam shape is Gaussian. The factor $2k/\lambda^2$ converts from flux units to temperature units using Rayleigh–Jeans approximation to the Planck blackbody law.

Note that in this approximation the entire radiometric system is characterized by two constants. For the case above they are in effect the value of the system temperature or the calibration noise tube in units of apparent brightness temperature and in units of apparent flux density. Scans or spectra can have either or both units on the ordinate. If the available calibration source (with known flux density) is not small compared with the antenna beam, then only one constant can be determined (!) and that constant corresponds to apparent brightness temperature units.

For an extended source

$$\int T_{B,\,Ap}(\Omega)\,d\Omega = (2k/\lambda^2)S_\nu, \tag{26}$$

and the scale for $T_{B,\,Ap}$ should be adjusted to make this relationship true. However the factor between the peak apparent brightness temperature and S_ν remains unknown (from an extended calibration source) unless the beam solid angle is measured some other way.

Other more complex cases can occur. For example, one might have a point source with unknown flux and also an extended source with known flux. Or one might have an extended source with known angular structure (such as the edge of the Sun or Moon). These cases are solvable, at least in principle, if one can obtain enough information to calculate the beam shape. The point is, however, that these special cases are equally difficult to deal with regardless of whether or not one knows the value of the calibration noise tube in units of antenna temperature.

B. CALIBRATION: $(S\text{-}R)/R$ vs. $(S\text{-}R)/C$

The result of measuring a spectrum in radio astronomy is usually a plot with apparent flux units and/or apparent brightness temperature units on the ordinate and frequency or Doppler velocity on the abscissa. In this section we assume that either a calibration noise tube or the system temperature is known in brightness temperature units and/or flux units, and we consider the problem of transferring these units to the ordinates of the spectra. In Section III, A we considered the problem of obtaining these values for the calibration noise tube and system temperature and in Section III, D, the problem of abscissae for the spectra will be discussed.

If we imagine a spectral feature whose strength is constant but whose frequency moves about within the bandpass of the spectrometer, then a proper calibration procedure should produce the same amplitude for this feature regardless of where in the bandpass it falls. In addition to any amplitude variations with frequency in the spectrometer itself, that portion of the receiver prior to the spectrometer contains filters which shape the bandpass. The calibration procedure, then, must compensate for all these spectral variations.

There are two popular calibration procedures, referred to as $(S\text{-}R)/R$ and $(S\text{-}R)/C$, based on two different assumptions about the behavior of the front end of the receiver. The "$S\text{-}R$" in these pseudoformulas refers to "signal minus reference" in the switching scheme as discussed in Section III, C below. The divisor (R or C) is discussed in this section.

If one has available a calibration noise source that (a) has a known value in flux units and/or brightness temperature units, and (b) is constant as a function of both frequency and time, then one can use the $(S\text{-}R)/C$ scheme. One simply turns on the calibration noise source and records the response of the spectrometer. This procedure must be repeated often enough to insure that temporal drifts in the receiver are compensated for, and must be done for a time long enough to insure that the noise in the calibration process contributes a negligible error to the final spectrum. The calibration noise source can be turned on during the actual observations, in which case it adds somewhat to the system temperature; or the observations can be stopped while the calibration procedure is carried out as a separate step, in which case it detracts somewhat from the available integration time. If the calibration procedure is carried out during the reference part of the switching cycle, then noise is not added to the spectrum nor is time wasted, however, this scheme cannot be used with gain modulation nor in certain other special cases.

The rms gain error ΔG due to noise in the $(S\text{-}R)/C$ calibration procedure can be calculated from

$$\left(\frac{\Delta G}{G}\right)^2 = \frac{2\rho^2 + 2\rho + 1}{\beta t} \tag{27}$$

where ρ is the ratio of system temperature to calibration temperature (usually but not necessarily >1), β is the resolution width, and t is the length of time the calibration noise source is on. It is assumed to be off for an equal time to provide a comparison.

The effect of gain errors is to add extra noise to the peak of strong features or to any region of the spectrum with a large offset such as might be due to a sloping or curved baseline. The fractional gain error $\Delta G/G$ should be made negligibly small compared with the ordinary noise. This usually means making $\Delta G/G$ small compared with the fractional noise on the strongest feature. Figure 10 is a graphical version of Eq. (27).

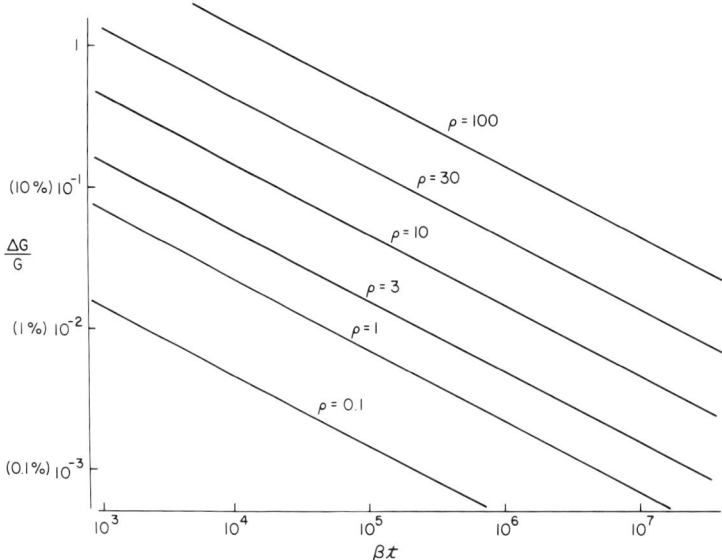

FIG. 10. Fractional gain error due to noise with the $(S\text{-}R)/C$ calibration procedure.

For the ordinate of the final spectrum in the $(S\text{-}R)/C$ scheme, one calculates

$$\frac{P_S(\nu) - P_R(\nu)}{P_{\text{con}}(\nu) - P_{\text{coff}}(\nu)} \times T_{\text{cal}}, \qquad (28)$$

where $P_S(\nu)$ is the raw (normalized) spectrum taken on the source ("signal"), $P_R(\nu)$ is taken off the source ("reference"), $P_{\text{con}}(\nu)$ is taken with the calibration noise source on, $P_{\text{coff}}(\nu)$ is taken with it off, and T_{cal} is the value of the calibration noise source in whatever units are wanted on the final spectrum. Any normalization constant must be the same for the four P's, but is otherwise arbitrary.

The denominator in Eq. (28) represents the response of the system to a flat spectrum, so by dividing by it, one equalizes the response across the bandpass.

Alternatively, if the system temperature (a) has a known value in flux units and/or brightness temperature units, and (b) is constant as a function of both frequency and time, then one can use the $(S\text{-}R)/R$ scheme. This scheme is popular with autocorrelation spectrometers because the formulas occur in a natural way and because these assumptions about the system temperature are a good approximation for the amplifier front ends at centimeter wavelengths usually used with correlators. Furthermore, the $(S\text{-}R)/C$ scheme cannot be used with one-bit digital autocorrelators because a one-bit system is insensitive to added signals with flat spectra such as the added noise of the calibration noise source.

For the ordinate of the final spectrum in the $(S\text{-}R)/R$ scheme, one calculates

$$\frac{P_S(\nu) - P_R(\nu)}{P_R(\nu)} \times T_{\text{sys}} \qquad (29)$$

where T_{sys} is the system temperature. Because of our assumptions about the system temperature, the denominator in this equation—the bandpass noise spectrum—also represents the response of the system to a flat spectrum.

The fractional gain error due to noise in the $(S\text{-}R)/R$ calibration scheme is just

$$\frac{\Delta G}{G} = \frac{\Delta P_R(\nu)}{P_R(\nu)}, \qquad (30)$$

and this gain error will have a negligibly small effect on the final spectrum unless the signal or offset is comparable to T_{sys}. It can be shown that the excess noise from this cause is

$$\sigma^1/\sigma = (q^2/2 + q + 1)^{1/2}, \qquad (31)$$

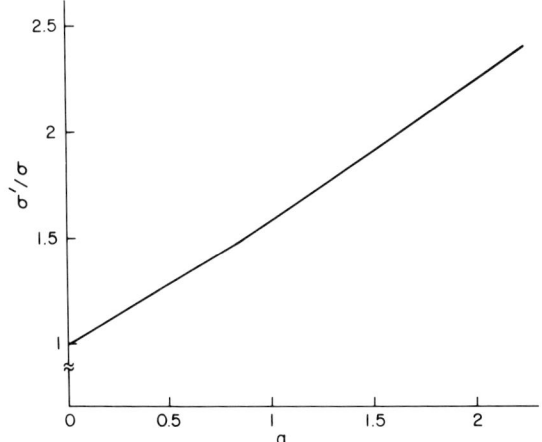

FIG. 11. Excess noise on a strong feature with the $(S-R)/R$ calibration procedure. q is the ratio of the strength of the feature to T_{sys}.

where q is the strength of a feature as a fraction of T_{sys}, σ^1 is the actual dispersion, and σ is the dispersion for small q (e.g., baseline regions) as calculated from Eq. (32) in Section III, C. Figure 11 is a graphical version of Eq. (31).

An intermediate situation also frequently occurs in which one assumes the system temperature to have a flat spectrum but not a constant value in time. This case can be handled by comparing the system temperature with a calibration noise source as frequently as necessary, and then using the time variable or time averaged T_{sys} in the preceding equations.

C. Switching Techniques and Baseline Curvature

The overall shape of the noise spectrum at the input to the spectrometer is largely determined by the bandpass shape of various amplifiers and filters in the receiver. The actual signal to be observed usually appears as a small perturbation on top of this shape. Thus some sort of switching scheme is usually necessary. One differences two spectra taken with ("signal") and without ("reference") the signal present to obtain the final spectrum. This procedure should yield a flat baseline for the difference spectrum.

However, one of the most persistent and difficult problems in spectral-line radio astronomy is the problem of obtaining good baselines. In a surprising number of observations, the limit on the detail that can be seen in a spectrum or the limit on the detectability of weak features is determined not by noise but by instrumental baseline curvature. The various sources of instrumental baseline curvature can be divided into two categories—those that produce curvature by varying in time, and those that are independent of time. The problem of ameliorating baseline curvature of time-independent origin is essentially the problem of choosing a suitable reference spectrum to switch against. And the problem of ameliorating baseline curvature of time-dependent origin involves choosing the rate of switching. The switching cycle should be chosen to be short compared with the time scale for significant changes in the relevant system parameters but long compared with the dead time or blanking time involved in the switch. This time scale ranges from milliseconds to hours depending on the details of the design of both the front end and the spectrometer.

A one-bit correlation system is insensitive to system gain (although sensitive to spectral shape, of course) and most radio astronomy front ends exhibit more rapid and more dramatic changes in gain than in other parameters such as system temperature. A gain-stabilized ("automatic gain control") filter spectrometer can also be built. With a system insensitive to gain changes, the switching cycle can usually be much longer. With such long cycles, several new possibilities occur; in particular the antenna can be moved between the signal and reference measurements and one reference spectrum can be used with several signal spectra.

The perfect switching scheme would be one in which the signal component in the spectrum is switched off and on without any other changes. Thus the difference between "signal" and "reference" would be precisely the quantity to be measured. Various practical switching schemes can be compared against this ideal. A more detailed discussion of switching schemes and baselines is given by Ball (1976), and baseline fitting procedures will be discussed in Section IV, A.

The noise equation in the form usually used in radio astronomy is

$$\Delta T = 2\gamma T_{\text{sys}}/(\beta t)^{1/2} \tag{32}$$

In this equation ΔT is the rms noise on the spectrum, T_{sys} is the system temperature in whatever units are wanted for ΔT, β is the noise width of the resolution, and t is the integration time.

It can be shown that ΔT is minimized by dividing a given observing session into equal integration times for signal and reference. The 2 in the numerator is the price one pays for this switching and is the product of two $\sqrt{2}$'s, one resulting from spending only half the time looking at the signal and another $\sqrt{2}$ from differencing two equally noisy spectra. The γ in the numerator represents the clipping correction for a clipped correlation system and/or any additional factors resulting from other observing modes. γ will be unity for a filter spectrometer if the signal and reference spectra contain equal integration time.

In practice, not all spectrometers behave according to this relation. In particular, filter receivers without gain stabilization often exhibit a lower limit on ΔT regardless of how long t is made. It may be useful to plot the measured ΔT against t on a log-log scale to look for departures from a straight line with a slope of $-\frac{1}{2}$. Such behavior is usually related to gain fluctuations in the receiver or spectrometer and can be ameliorated by a carefully designed automatic gain control (AGC). And, of course, Eq. (32) has nothing to say about baseline curvature or other instrumental effects.

D. Doppler Velocities and the Scale for the Abscissa

The abscissa for spectra in radio astronomy can be frequency, wavelength, or Doppler velocity referred to some rest frequency, and the velocity reference frame can be the Sun, the local standard of rest (discussed below), or some other. Since frequency is usually measured directly, it might seem to be the logical quantity, but frequency is usually inconvenient for the abscissa, especially when one is comparing spectra of various species toward a given source.

Instead, radio astronomers traditionally use Doppler velocities referred to the local standard of rest—a velocity reference frame moving approximately with the average velocity of all the stars in the solar vicinity. By convention, the velocity of the Sun with respect to the local standard of rest is taken to be 20 km/sec toward $\alpha = 18^\text{H}$, $\delta = +30°$ (1900).

Radio astronomers usually convert from frequency offsets to Doppler velocities using

$$\Delta \nu / \nu_0 \simeq -v/c \qquad (33)$$

where $\Delta \nu$ is the change in frequency due to the Doppler effect, ν_0 is the rest frequency for the particular line observed, v is the component of the velocity along the line of sight, and c is the speed of light. By convention a positive v corresponds to increasing distance between source and observer, and this convention causes the minus sign. The approximation in this equation is reasonable provided $|\Delta \nu / \nu_0| \ll 1$, i.e., provided relativistic effects do not need to be taken into account.

Optical astronomers traditionally use wavelengths referred to the Sun for the abscissae of their spectra. This is appropriate because their spectrometers are usually calibrated in wavelengths and because many lines from many species are frequently present in their spectra simultaneously.

Optical astronomers sometimes also use Doppler velocities referred to the Sun, or z, defined as

$$z = \Delta\lambda/\lambda_0 \simeq v/c, \qquad (34)$$

where $\Delta\lambda$ is the change in wavelength due to the Doppler effect, and λ_0 is the rest wavelength for the particular line observed. The approximation in this equation is reasonable provided $z \ll 1$.

If v is so large that relativistic effects need to be taken into account, then neither of these equations gives the correct v and they also do not agree with each other. Furthermore, in the relativistic case, it is not correct to add up several velocities as if they were scalars and then apply the relativistic formula to the scalar sum. Fortunately, relativistic velocities on abscissae of spectra occur rarely in radio astronomy.

Astronomical spectra are taken from a moving platform, usually the Earth, and so must have their abscissae corrected for the motion of the observer with respect to some velocity reference frame. The largest components of the Earth's motion are the revolution, which amounts to almost 30 km/sec, and the rotation, which can be almost 0.5 km/sec. Computer programs that calculate these velocities are available (e.g., Ball, 1969a). Usually one uses such a program during data acquisition to adjust the local oscillator frequency so as to produce spectra with a specified range of Doppler velocities referred to the local standard of rest. Such spectra then can be directly compared or averaged together even if they are made at different times, i.e., with different contributions from the Earth's motion.

IV. Selected Problems in Interpretation of Spectra

A. Fitting Gaussians and Baselines

Sometimes it is desirable to subtract a polynomial or other smooth curve from the spectrum to remove baseline curvature, at least in part. The usual procedure involves fitting a polynomial in a least-squares sense to those parts of the spectrum on which no spectral features are present, and then subtracting the polynomial from the whole spectrum. The idea is that instrumental baseline curvature is usually smooth and can be approximated by the first few terms of its Taylor-series expansion. However, extreme caution is recommended when using this procedure with higher order polynomials because such polynomials can either create or destroy spectral features.

Sometimes it is desirable to fit the spectrum in a least-squares sense with one or more functions, such as Gaussians, to obtain parameters for comparison with a theoretical model or with other spectra. Although any spectrum can be fitted by the sum of a number of Gaussians, this procedure is not very useful unless the number of Gaussians is small or unless one is working with a theoretical model that predicts Gaussians.

So the motivations for fitting polynomials and for fitting Gaussians are different, but the two procedures are discussed together in this section because the mathematics involved are similar and because it is sometimes desirable to do the fittings simultaneously.

The discussion of fitting procedures in this section is largely from the work of Kaper et al. (1966). Several computer programs now exist for fitting Gaussians or Gaussians and polynomials simultaneously based on this analysis by Kaper et al. (e.g., Cesarsky, 1970; Ball et al., 1970).

The data are thought of as a set y_i with associated x_i. The goal of the fitting procedure is to find some function $F(\mathbf{p}, x)$ such that

$$S(\mathbf{p}) \equiv \sum_{i=1}^{n} [F(\mathbf{p}, x_i) - y_i]^2 \qquad (35)$$

is minimized for variations of each of a set of parameters p_l characterizing F. S is the sum of the squares of the residuals in the fit.

We differentiate S with respect to each of the p_l and set the result to zero. This gives the normal equations,

$$\sum_i [F(\mathbf{p}, x_i) - y_i] \frac{\partial F(\mathbf{p}, x_i)}{\partial p_l} = 0; \qquad l = 1, m. \qquad (36)$$

If F is a linear function of the p_l, as will be the case, for example, with any set of functions such as polynomials to be added together to give F with only indeterminate amplitudes, then the normal equations are m linear equations in m unknowns and can be solved directly with standard matrix inversion techniques. If, however, F is a nonlinear function of the p_l, as will be the case, for example, with any set of functions such as Gaussians that contain p_l intrinsically, then the normal equations are m nonlinear equations in m unknowns and can be solved only by some iteration scheme. Furthermore, in the nonlinear case, multiple solutions sometimes exist. That is, the surface defined by $S(\mathbf{p})$ sometimes has several local minima. Iterative solutions to such problems find the minimum nearest to some initial guess.

A simple procedure for fitting polynomial baselines utilizes three FORTRAN subprograms supplied in the IBM Scientific Subroutine Package (1970). APCH sets up the normal equations for a least-squares fit of the data with Chebyshev polynomials of specified degree; APFS computes the solutions to the normal equations to get the coefficients of the Chebyshev polynomials; and CNPS computes the value of the series expansion in Chebyshev polynomials to subtract from each point.

A way to deal with the nonlinear case is to take only the first two terms of the Taylor-series expansion of $F(\mathbf{p}, x)$ around a particular $\mathbf{p}^\circ$—an initial guess. That is, we approximate

$$F(\mathbf{p}, x) \simeq F(\mathbf{p}^\circ, x) + \sum_l \left.\frac{\partial F(\mathbf{p}, x)}{\partial p_l}\right|_{\mathbf{p}^\circ} (p_l - p_l^\circ). \tag{37}$$

Then the linearized normal equations are

$$\sum_i [y_i - F(\mathbf{p}^\circ, x_i)] \left.\frac{\partial F(\mathbf{p}, x_i)}{\partial p_l}\right|_{\mathbf{p}^\circ} = \sum_k \Delta p_k \sum_i \left(\left.\frac{\partial F(\mathbf{p}, x_i)}{\partial p_k}\right|_{\mathbf{p}^\circ} \left.\frac{\partial F(\mathbf{p}, x_i)}{\partial p_l}\right|_{\mathbf{p}^\circ} \right.$$
$$\left. + \left.\frac{\partial^2 F(\mathbf{p}, x_i)}{\partial p_k \partial p_l}\right|_{\mathbf{p}^\circ} [y_i - F(\mathbf{p}^\circ, x_i)] \right); \; l = 1, m. \tag{38}$$

These are now m linear equations in m unknowns and can be solved directly with standard matrix inversion techniques to give $\Delta p_l = p_l - p_l^\circ$. Then the procedure can be iterated using a new guess, $p_l = p_l^\circ + \Delta p_l$, etc.

For most practical cases, the second-derivative terms are zero except on the diagonal ($k = l$). Moreover, Kaper et al. argue that even these diagonal terms are usually small and can be neglected. For the case that the x_i are spaced at the Nyquist interval or closer, Kaper et al. show how to evaluate those sums over i that do not depend on y_i by replacing them by integrals evaluated algebraically. This procedure results in a more complex program, but certainly saves computer time. After the linearized normal equations are set up, a subprogram such as GELS from the IBM Scientific Subroutine Package (1970) can be used to compute the Δp_l for each iteration.

Using the linearized normal equations represents a multidimensional generalization of Newton's method for finding roots of nonlinear equations. Near an initial guess, one approximates the function by a straight line and extrapolates the line to zero to give an improved guess.

Untoward behavior occurs when the function F is such that near $\mathbf{p}^\circ$ the higher order derivatives are not small. Specifically, the straight-line approximation may overshoot the minimum. It can be shown, however, that the direction of Δp_l is correct, so one can try proceeding in this direction by a smaller step. If S increases, one can try $E\Delta p_l$ where $E < 1$.

Even if the iteration procedure works well, it may turn out that some of the p_l are poorly determined, i.e., have large errors associated with them. Kaper *et al.* discuss how to estimate the errors in p_l using linear error propagation.

However, in some simple cases the expected errors are easy to estimate. Suppose one is fitting a single Gaussian to an isolated feature. For this case it can be shown that

$$\Delta A \simeq \Delta T (\beta/W)^{1/2}, \qquad (39)$$

where ΔA is the expected dispersion in the amplitude of the Gaussian, ΔT is the noise dispersion on the spectrum (or in the residuals in the fit), β is the resolution, and W is the width of the feature. Note that this ΔA is just what one would have calculated from Eq. (32) with W in place of β. Thus ΔA is independent of β, provided β is narrow enough not to smear W. It is sometimes erroneously believed that narrowing β increases ΔA.

Then

$$\Delta W/W \simeq \Delta A/A, \qquad (40)$$

that is, the expected percentage dispersion in W is the same as in A. Finally

$$\Delta P \simeq \Delta W/2, \qquad (41)$$

where ΔP is the expected dispersion in the position of the center of the Gaussian. If we think of the two half-power points determined separately with some dispersion, then W is the difference between them and so has $\sqrt{2}$ more dispersion, and P is the average of the two and so has $\sqrt{2}$ less dispersion, which leads to the 2 in Eq. (41).

A typical fitting problem might involve a spectrum with several emission or absorption features somewhat overlapping in the center of the spectrum and with baseline regions (with no features present) on the edges of the spectrum. Two alternatives are (a) to fit the baseline regions (only) with a polynomial, subtract it from the whole spectrum, and then fit the spectral features with Gaussians as a separate step, or (b) to fit the entire spectrum simultaneously with the sum of a polynomial and Gaussians. The simultaneous fit (b) is usually best, but the choice is not as clear-cut as sometimes supposed. In particular, if there is insufficient baseline to fit a polynomial in scheme (a), then scheme (b) usually fails also because there is no way for the program to distinguish between signal and baseline. Alternatively, if there are adequate amounts of baseline available, then the two schemes usually give about the same answer.

It sometimes happens that one or more of the p_l are to be constrained in some way. If a particular p_l is to be constant, for example, it can be removed from the fitting procedure altogether. It may happen that two or more of the p_l are to be dependent on each other, for example two or more Gaussians are to have constant spacing. This occurs when a molecule has known hyperfine structure within the passband. This problem again should be handled by removing the parameters from the matrix and replacing them with a single parameter with derivatives equal to the sum of the derivatives of the removed parameters. The alternative of setting derivatives to zero or equal does not work.

B. Polarization—Stokes Parameters

The Stokes parameters represent one of several alternative schemes for describing the polarization properties of electromagnetic radiation. Stokes parameters may be used for either continuum sources or for spectra, of course, but only in maser emission sources and in pulsars have strong linear and circular polarization been found.

There are four Stokes parameters and it can be shown that four quantities are the minimum necessary to measure and also to describe the complete polarization properties of radiation at each point across the spectrum. We need a minimum of four spectra at different polarizations to calculate the Stokes parameters.

The following definitions of the Stokes parameters are those given by Born and Wolf (1964) except that the radio definition of the sense of circular polarization is used rather than the optical one. Each of the Stokes parameters is a function of frequency or Doppler velocity, and also of two angles on the sky, however, these dependencies are omitted in the notation for simplicity.

The four Stokes parameters are

$$S_0 = I_R + I_L = I(\theta) + I(\theta + 90°), \quad \text{any } \theta, \tag{42}$$
$$S_1 = I(0) - I(90°), \tag{43}$$
$$S_2 = I(45°) - I(135°), \tag{44}$$
$$S_3 = I_R - I_L, \tag{45}$$

where I_R and I_L are the specific intensities (or brightnesses) measured in right and left circular polarization, and $I(\theta)$ is the specific intensity measured in linear polarization at the polarization position angle θ. For a small-diameter source, these definitions can be replaced by equivalent definitions in terms of flux densities.

Right circular polarization corresponds to clockwise rotation of the electric vector as viewed along the direction of propagation, so the electric vector traces a right-handed screw thread. This is the standard radio definition, but is opposite to the optical definition. The polarization position angle for linear polarization is measured, positive counterclockwise, from zero at north–south on the sky. Values of θ differing by 180° are equivalent except possibly for instrumental effects.

Incoherent addition of intensities, e.g., from two sources, gives Stokes parameters that are just the sum of the Stokes parameters from the separate sources.

The specific intensity measured at an arbitrary polarization position angle θ is

$$I(\theta) = \tfrac{1}{2}S_0 + \tfrac{1}{2}(S_1^2 + S_2^2)^{1/2}\cos(2\theta - 2\phi), \tag{46}$$

where

$$2\phi = \tan^{-1}(S_2/S_1), \tag{47}$$

and ϕ is the position angle of the major axis of the polarization ellipse. This relation can be used to determine S_1 and S_2 from data taken at arbitrary polarization angles. Equation (46) above can be used to fit these data in a least squares sense to determine S_0, $(S_1^2 + S_2^2)^{1/2}$, and ϕ (see Ball, 1969b). Then

$$S_1 = (S_1^2 + S_2^2)^{1/2}\cos 2\phi, \tag{48}$$
$$S_2 = (S_1^2 + S_2^2)^{1/2}\sin 2\phi. \tag{49}$$

Other polarization parameters can be calculated from the Stokes parameters. The fractional linear polarization is

$$p_L = (S_1^2 + S_2^2)^{1/2}/S_0 \tag{50}$$

And the fractional circular polarization is

$$p_C = \frac{S_3}{S_0} = \frac{I_R - I_L}{I_R + I_L} \tag{51}$$

So a positive p_C represents dominantly right circular polarization and a negative p_C left. The (total) fractional polarization is

$$p = (p_L^2 + p_C^2)^{1/2} = (S_1^2 + S_2^2 + S_3^2)^{1/2}/S_0 \tag{52}$$

The ellipticity angle χ can be obtained from

$$\sin 2\chi = p_C/p = S_3/(S_1^2 + S_2^2 + S_3^2)^{1/2} \tag{53}$$

The angle 2ϕ is the longitude and the angle 2χ is the latitude on the Poincaré sphere. Figure 12 is a sketch of the three-dimensional Poincaré sphere and is useful in visualizing these relations. The fractional polarization p is represented by the fraction of the radius of the circle inside P, the fractional linear polarization p_L by the fraction of the radius inside the projection of P onto the equatorial plane, and the fractional circular polarization p_C by the fraction of the radius inside the projection of P onto the vertical axis.

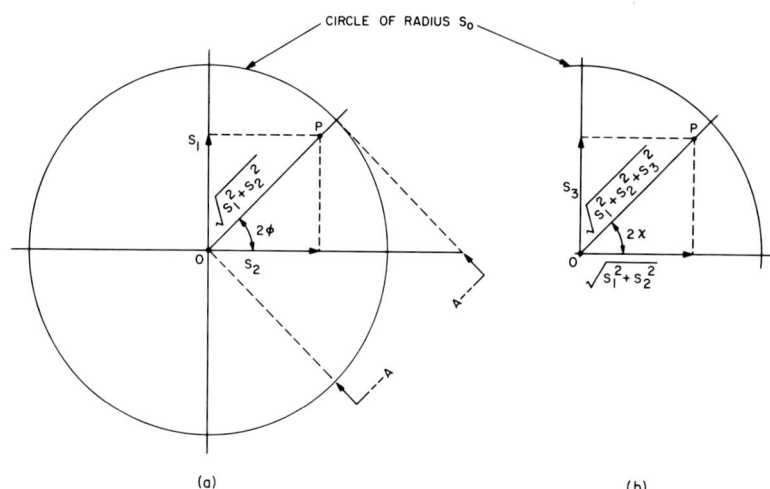

FIG. 12. Three-dimensional Poincaré sphere: (a) top view; (b) side view [section A-A in (a)].

Elliptical polarization results from the incoherent addition of linearly and circularly polarized components, however, the fractional polarization p must then be less than unity. Incoherent addition of radiation with pure linear, left, or right circular polarization yields a mixture that must satisfy

$$p \leq p_L + |p_C| \leq 1. \tag{54}$$

If $p_L + |p_C| > 1$, then the radiation cannot be made in this way.

It is usually preferable to plot and work with spectra of the Stokes parameters rather than, for example, p_L, p_C, and χ, because the error limits for these derived quantities are not usually constant across a spectrum. Plots of ϕ and χ are sometimes useful but difficult to interpret even if the error limits are plotted also.

C. Problems of Multiple Components

If the opacity is small then an observed spectrum is the convolution of the Doppler velocity distribution of the molecules (or atoms) in the source by the spectrum of the molecules themselves. If the molecular spectrum can be characterized as a single line with either zero width or a known width, then inverting this convolution is easy. It sometimes happens, however, that hyperfine components in a molecule are close enough together in frequency to be blended by the Doppler velocity spread in the source. Interpreting spectra in this case is more difficult. This section is a discussion of this problem.

Suppose that $P_0(\nu)$ is the observed spectrum (near $\nu = \nu_0$, the line rest frequency), $S(\nu)$ is the source Doppler frequency distribution (nonzero around $\nu = 0$), and $P_L(\nu)$ is the molecular spectrum (nonzero around $\nu = \nu_0$) as measured, for example, in the laboratory. Then

$$P_0(\nu) = \int_{-\infty}^{\infty} S(\nu - \nu^1) P_L(\nu^1) \, d\nu^1, \tag{55}$$

is an approximation for small opacity. (This equation is correct only for positive ν.)

We want to invert this convolution to obtain $S(\nu)$ given $P_0(\nu)$ and $P_L(\nu)$. To use the convolution theorem, we transform to lag space [cf. Eq. (3) in Section II, A],

$$R_0(\tau) = \int_{-\infty}^{\infty} P_0(\nu) \exp(2\pi j \nu \tau) \, d\nu, \tag{56}$$

and use a similar equation for $R_S(\tau)$ and $R_L(\tau)$. Then the convolution theorem gives

$$R_0(\tau) = R_S(\tau) R_L(\tau). \tag{57}$$

This equation can be inverted,

$$R_S(\tau) = R_0(\tau)/R_L(\tau), \tag{58}$$

whenever $R_L(\tau)$ is not zero. Then

$$S(\nu) = \int_{-\infty}^{\infty} R_S(\tau) \exp(-2\pi j \nu \tau) \, d\tau, \tag{59}$$

is the formal solution to this problem.

It is also possible to invert the convolution in Eq. (55) by manipulations in frequency space. Specifically, one would convolve $P_0(\nu)$ by the Fourier transform of $1/R_L(\tau)$. This alternative procedure proves to be more difficult in practice.

The convolution theorem is true regardless of how the spectra are obtained, but with an autocorrelation system the Fourier transform of the observed spectrum—the autocorrelation function—may already be known, thus saving one Fourier transform.

In order to apply these ideas in practice to overlapping spectra, one needs to know or assume the spectrum of the molecule. The laboratory line rest frequencies are usually known as are the molecular line strengths. And the natural linewidths are usually negligible. However, translating the relative molecular line strengths into observed relative intensities proves to be difficult because of (a) the possibility of nonequi-

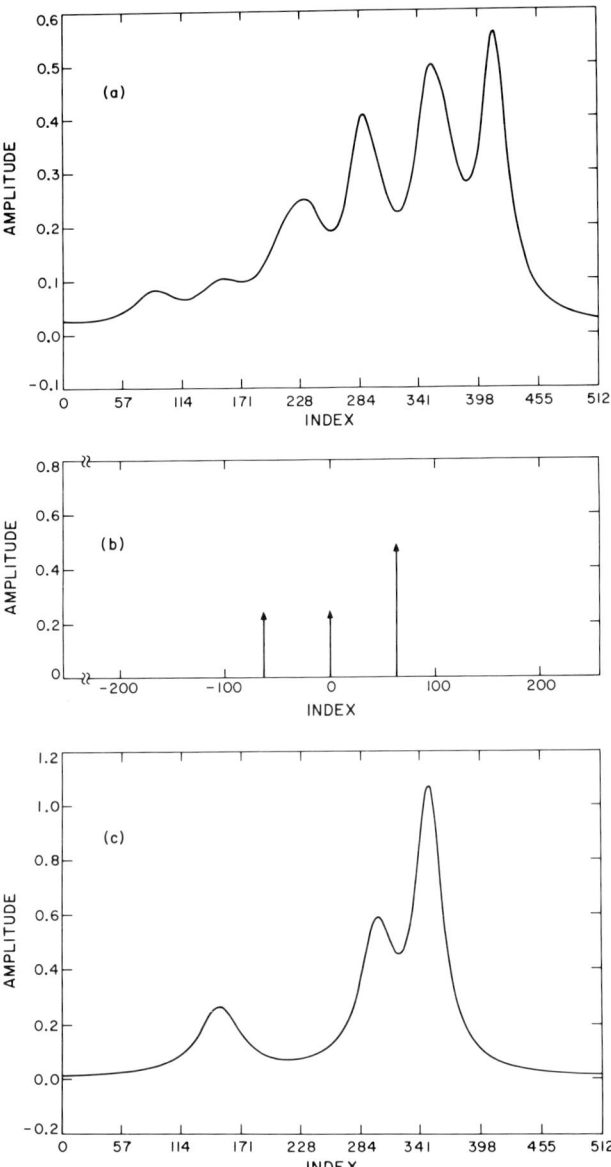

FIG. 13. Three calculated spectra: (a) represents the observed spectrum and is the convolution of the assumed laboratory spectrum (b) by the assumed source Doppler velocity distribution (c), which is made up of three peaks of the form $1/(1+x^2)$. The index numbers on the abscissae are arbitrary but represent frequencies.

librium populations, and (b) an unknown, but possibly large, opacity. In general, the observed line shape is a complex combination of the effects of kinematics and radiative transfer, and there is no way to separate the two. We discuss in this section one particularly simple case that can be solved—the case in which the molecular spectrum can be approximated by a finite number of delta functions of known frequency and known strength.

If the molecular spectrum is

$$P_L(\nu) = \sum_k C_k \delta(\nu - \nu_k), \qquad (60)$$

then

$$R_L(\tau) = \sum_k C_k \exp(2\pi j \nu_k \tau). \qquad (61)$$

Certain other functional forms for $P_L(\nu)$, such as Gaussians of specified width, are almost equally easy to work with. $R_L(\tau)$ in Eq. (61) will be zero for certain τ only in special cases involving some sort of symmetry. There is no restriction on the number of components (the range of k in the sum); indeed, the more components one has, the less likely one is to encounter zeros in $R_L(\tau)$.

As an example of the use of this technique, Fig. 13a shows the result of convolving an assumed laboratory spectrum (Fig. 13b) by an assumed source Doppler frequency distribution (Fig. 13c). The observed spectrum (Fig. 13a) is difficult to interpret by eye because it contains only six obvious features and because it seems at first to suggest that the relative heights of the hyperfine components might not be as shown in Fig. 13b. Nevertheless, if the Fourier transform of the spectrum in Fig. 13a is divided by the Fourier transform of the laboratory spectrum (Fig. 13b) and the result transformed back into frequency space, then one recovers Fig. 13c exactly.

This process insures that Fig. 13c convolved by Fig. 13b gives Fig. 13a and the inversion is unique—no other form for Fig. 13c will do. The arbitrariness is in the choice of the relative strengths for the hyperfine components (Fig. 13b).

This technique is new and has not been used much in practice, but it shows considerable promise for the future as increasingly complex molecular spectra are observed.

Acknowledgments

I thank Norman Brenner for help with Fourier transforms and such, my colleagues at the Center for Astrophysics for many stimulating discussions and other help, and particularly Prof. A. E. Lilley for providing the stimulating environment in which this work was carried out. Financial support came from the U.S. National Science Foundation through grant GP-19717.

REFERENCES

ABLES, J. G. (1974). *Astron. & Astrophys., Suppl.* **15,** 383.
BALL, J. A. (1969a). "Some FORTRAN Subprograms used in Astronomy," Tech. Note 1969–42. Lincoln Lab., M.I.T., Cambridge, Massachusetts.
BALL, J. A. (1969b). "Observations of OH Radio Emission Sources," Ph.D. Thesis, Harvard University, and Tech. Rep. 458. Lincoln Lab., M.I.T., Cambridge, Massachusetts.
BALL, J. A. (1972). "The Maximum Entropy Method of Spectral Analysis as Applied to Spectral-Line Radio Astronomy," Internal Memo. Harvard College Observatory, Cambridge, Massachusetts.
BALL, J. A. (1973). *IEEE Trans. Instrum. Meas.* **22,** 193.
BALL, J. A. (1976). *Methods Exp. Phys.* **12c** (to be published).
BALL, J. A., CESARSKY, D., DUPREE, A. K., GOLDBERG, L., and LILLEY, A. E. (1970). *Astrophys. J.* **162,** L25.
BLACKMAN, R. B., and TUKEY, J. W. (1958). "The Measurement of Power Spectra." Dover, New York.
BORN, M., and WOLF, E. (1964). "Principles of Optics." Pergamon, Oxford.
BOWERS, F. K., and KLINGER, R. J. (1974). *Astron. & Astrophys., Suppl.* **15,** 373.
BOWERS, F. K., WHYTE, D. A., LANDECKER, T. L., and KLINGER, R. J. (1973). *Proc. IEEE* **61,** 1339.
BRACEWELL, R. N. (1965). "The Fourier Transform and Its Applications." McGraw-Hill, New York.
BRENNER, N. M. (1973). "FORS1" (Fourier transform of real symmetric data) (privately circulated memorandum).
BURNS, W. R., and YAO, S. (1969). *Radio Sci.* **4,** 431.
CESARSKY, D. (1970). "Fitting Gaussians to Spectra," Internal Memo. Harvard College Observatory, Cambridge, Massachusetts.
CLARK, B. G. (1970). *Annu. Rev. Astron. Astrophys.* **8,** 115.
COLE, T. W. (1968a). *Opt. Technol.* **1,** 31.
COLE, T. W. (1968b). *Aust. J. Phys.* **21,** 273.
COLE, T. W. (1973a). *Proc. IEEE* **61,** 1321.
COLE, T. W. (1973b). *Astrophys. Lett.* **15,** 59.
COLE, T. W., and ABLES, J. G. (1975). *Astron. & Astrophys.* (Submitted for publication).
COOPER, B. F. C. (1970). *Aust. J. Phys.* **23,** 521.
GORDON, E. L. (1966). *Proc. IEEE* **54,** 1391.
HAGEN, J. B., and FARLEY, D. T. (1973). *Radio Sci.* **8,** 775.
HECHT, D. L. (1973). "The Development of an Acousto-Optic Spectrum Analyzer," Tech. Rep. Applied Technology Division of Itek Corporation, Sunnyvale, California.
IBM Scientific Subroutine Package, (1970). "Programmer's Manual," IBM Publ. GH20-0205, 5th Ed. IBM, White Plains, New York.
KAPER, H. G., SMITS, D. W., SCHWARZ, U., TAKAKUBO, K., and VAN WOERDEN, H. (1966). *Bull. Astron. Inst. Neth.* **18,** 465.
LACOSS, R. T. (1971). *Geophysics* **36,** 661.
LEVINSON, N. (1947). *J. Math. Phys.* **25,** 261. (See also WEINER, N. (1964). p. 136 ff.)

RICE, S. O. (1954). *In* "Selected Papers on Noise and Stochastic Processes" (N. Wax, ed.), p. 133. Dover, New York.

ROGERS, A. E. E. (1974). "Maximum Entropy Method for Unequally-Spaced Data in Two Dimensions and Numerical Tests of its Effectiveness," Tech. Memo. Haystack Observatory, Westford, Massachusetts.

VAN VLECK, J. H. (1943). "The Spectrum of Clipped Noise," Rep. No. 51. Radio Res. Lab., Harvard University, Cambridge, Massachusetts.

WEINER, N. (1964). "Time Series." M.I.T. Press, Cambridge, Massachusetts.

WEINREB, S. (1963). "A Digital Spectral Analysis Technique and its Application to Radio Astronomy," Ph.D. Thesis and R.L.E. Tech. Rep. No. 412. M.I.T., Cambridge, Massachusetts.

YEN, J. L. (1974). *Astron. & Astrophys., Suppl.* **15,** 483.

Author Index

Numbers in italics refer to the pages on which the complete references are listed.

A

Ables, J. G., 118, *126*, 169, *173*, 188, 189, 195, *218*
Abramowitz, M., 150, 168, *173*
Aimette, A., 12, *52*
Albershiem, W. J., 68, *128*
Alsup, J. M., 72, *129*
Altenhoff, W., 147, *173*
Altschuler, M. D., 48, *53*
Arm, M., 12, *52, 53*, 160, *174*
Armstrong, J. W., 117, *126*
Aubier-Giraud, M., 47, *63*

B

Baars, J. W. M., 172, *173*
Backer, D. C., 76, 87, 97, 98, 99, 100, 101, 102, 103, 121, *126*, *128*
Bagri, D. S., 172, *175*
Ball, J. A., 192, 195, 206, 208, 209, 213, *218*
Baldwin, J. E., 172, *173*
Batchelor, R. A., 109, 118, *126*, *127*
Baym, G., 125, *126*
Bell, S. J., 56, 57, 118, *128*
Bergland, G. D., 71, *126*
Biraud, Y., 168, *173*
Blackman, R. B., 94, 97, *127*, 182, *218*
Blum, E. J., 12, *52*
Boischot, A., 47, *53*
Boriakoff, V., 69, 76, 79, 106, 107, 109, 112, *126*, *127*
Born, M., 133, *173*, 212, *218*
Boynton, P. E., 121, 125, 126, *127*
Bowers, F. K., 172, *174*, 193, *218*
Bracewell, R. N., 14, *52*, 66, 91, 106, *127*, 132, 133, 141, 143, 149, 164, 166, 172, *173*, *175*, 180, 182, *218*
Brenner, N. M., 192, *218*
Briggs, B. H., 12, *52*, 117, *127*

Broderick, J. J., 76, 116, *127, 129*
Broten, N. W., 13, *52*
Brouw, W. N., 153, 154, 162, 166, *173, 174*
Burns, W. R., 58, 61, 62, *127*, 158, 168, *174*, 192, *218*

C

Cameron, A. G. W., 125, *127*
Campbell, D. B., 85, 87, 114, 115, *127, 128*
Carleton, N. P., 123, *128*
Carter, A. W. L., 18, *52*
Casse, J. L., 161, 172, *173, 174*
Cesarsky, D., 209, *218*
Chikayoshi, T., 14, *53*
Chiuderi, C., 3, *52*
Chiuderi Drago, F., 3, *52*
Christiansen, W. N., 5, 8, 13, 14, *52*, 132, 133, 155, 159, 172, *174*
Clark, B. G., 58, 61, 62, *127*, 190, *218*
Clark, R. R., 87, *127*
Clark, T. A., 76, 116, *127, 129*
Clemence, G. M., 122, *127*
Cohen, M. H., 132, *174*
Cole, T. W., 68, *127*, 160, *174*, 188, 189, 193, 197, *218*
Coles, W. A., 117, *126*
Collins, R. A., 56, 57, 118, *128*
Colvin, R. S., 172, *173*
Comlla, J. M., 57, 104, 114, 115, *127, 128*
Conway, R. G., 166, 172, *174*
Colley, J. W., 71, *127*, 164, *174*
Cooper, B. F. C., 193, *218*
Cordes, J. M., 94, 97, 101, *127, 128*
Costain, C. H., 172, *174*
Counselman, C. C. III., 79, 80, 114, 115, 119, 123, *127, 128*
Covington, A. E., 13, *52*
Craft, H. D., Jr., 69, 78, 79, 80, 81, 96, 98, 99, 101, 106, 114, 115, *127, 128*

221

AUTHOR INDEX

Cronyn, W. M., 112, 113, 116, *127*
Cudaback, D., 116, 124, *128*

D

D'Addario, L. R., 172, *173*
Daniels, G. R., 162, *174*
Darlington, S., 68, *128*
Davies, J. G., 62, 84, *127*
Davies, K., 86, *127*
Downs, G. S., 125, *128*
Drake, F. D., 80, 96, 98, *127*, *129*
Dulk, G. A., 51, *52*
Dupree, A. K., 209, *218*
Duthie, J. G., 123, *127*

E

Elsmore, B., 139, 162, *174*, *175*
Ekers, R. D., 171, *174*
Enome, S., 14, *53*
Erickson, G. R., 116, *129*
Erickson, W. C., 14, 15, *52*, 76, 116, *127*
Evans, J. V., *127*
Ewing, M. S., 109, *127*

F

Fainberg, J., 14, *52*
Farley, D. T., 193, *218*
Fejer, J. A., 113, *127*
Flammer, C., 168, *174*
Frater, R. H., 161, 162, *174*
Friefield, R. D., 109, *127*

G

Gabor, D., 11, *52*
Galt, J. A., 117, *127*
Goldberg, L., 209, *218*
Goldreich, P., 123, *127*
Gordon, E. L., 188, *218*
Graham, D. A., 121, *128*
Grebenkemper, C. J., 172, *173*
Greenstein, G. S., 125, *127*
Groth, J. E., 121, 125, 126, *127*
Gunn, J. E., 123, *128*

H

Hagen, J. B., 193, *218*
Hagfors, T., *127*

Hamaker, J. P., 162, 172, *173*, *174*
Hankins, T. H., 71, 72, 76, 77, 85, 91, 92, 93, 94, *127*, *128*, *129*
Hecht, D. L., 188, *218*
Heiles, C., 85, *127*
Helstrom, C. W., 65, *128*
Hemenway, P. D., 147, *174*
Hewish, A., 56, 57, 118, *128*, 132, *174*
Hills, R. E., 116, 124, *128*, 172, *174*
Hinder, R., 147, *174*
Hirsch, R. M., 87, *129*
Hoekema, T., 170, *174*
Högbom, J. A., 133, 155, 159, 170, 171, *174*
Hogg, D. E., 166, 172, *174*
Holmes, N., 12, *52*
Horowitz, P., 123, *128*
Horton, P. W., 84, *127*
Huguenin, G. R., 61, 76, 77, 87, 89, 90, 96, 97, 107, 113, 123, *128*, *129*
Hulse, R. A., 57, *128*
Hunt, G. C., 120, 123, *128*
Hutchinson, D. P., 121, 125, 126, *127*

I

Imrie, K. S., 161, *174*

J

Jackson, P. D., 14, *52*
Janssen, M. A., 172, *174*
Jenkins, G. M., 97, *128*
Jennings, J. E., 172, *173*
Julian, W. H., 123, *127*

K

Kai, K., 14, 31, 46, 49, 50, *52*, *53*
Kakinum, T., 5, 13, 14, *53*
Kaper, H. G., 209, *218*
Kellermann, K. I., 173, *174*
Kenderdine, S., 164, *174*
Klauder, J. R., 68, *128*
Klinger, R. J., 193, *218*
Knowles, S. H., 76, 116, *127*
Ko, H. C., 138, *174*
Kobayashi, S., 14, *53*
Krishnamohan, S., 118, *126*

AUTHOR INDEX

Kuiper, T. B. H., 76, 116, *127*
Kundu, M. R., 14, *52*

L

Labrum, N. R., 20, 35, 45, *52, 53*
Lacey, J. D., 172, *174*
Lacoss, R. T., 169, *174*, 195, *218*
Lambert, L. B., 12, *52, 53*, 160, *174*
Landau, H. J., 168, *174*
Landecker, T. L., 172, *174*, 193, *218*
Lang, K. R., 102, 113, 115, 117, *128*
Large, M. I., 57, 62, *127, 129*
LeBlanc, Y., 47, *53*
LePoole, R. S., 171, *174*
Lilley, A. E., 209, *218*
Little, A. G., 8, *53*
Levinson, N., 195, *218*
Lovelace, R. V. E., 58, 59, 98, 99, 101, *128*
Lyne, A. G., 84, 106, 117, 121, *127, 128*

M

MacDonald, G. H., 166, 172, *174*
McLean, D. J., 12, 36, 37, 38, 46, 48, *53*, 160, *174*
Manchester, R. N., 66, 76, 77, 81, 89, 90, 96, 107, 120, 123, 124, 125, *126, 128, 129*
Mathewson, D. S., 5, 6, 8, 13, 14, *52*
Mathur, N. C., 133, 166, *174, 175*
Miley, G. K., 171, *174*
Mills, B. Y., 8, *53*
Moreton, G. E., 47, *53*
Morimoto, M., 14, *53*
Murdin, P., 123, *127*

N

Nanos, G. P., Jr., 121, 125, 126, *127*
Nelson, J., 116, 124, *128*
Neville, A. C., 139, 173, *175*
Newkirk, G., 48, *53*
Noci, G., 3, *52*

O

O'Brien, P. A., 13, *53*
O'Meara, T. R., 74, *128*
Orsten, G. S. F., 69, *128*

Ostriker, J. P., 123, *128*
O'Sullivan, J. D., 161, 162, *174*

P

Pacini, F., 125, *129*
Papaliolios, C., 123, *128*
Partridge, R. B., 121, 125, 126, *127*
Pauliny-Toth, I. I. K., 173, *174*
Payne, R. R., 116, *129*
Payten, W. J., 20, *53*
Petengill, G. A., 119, 123, *128*
Peters, W. L., 120, 123, *128*
Pethick, C., 125, *126*
Phillips, G. J., 117, *127*
Pilkington, J. D. H., 56, 57, 118, *128*
Pines, D., 125, *126*
Pollak, H. O., 168, *174, 175*
Price, A. C., 68, *128*
Price, K. M., 172, *173*
Price, R. M., 76, 109, *127, 129*

R

Radhakrishnan, V., 125, *128*
Ramsey, H. E., 47, *53*
Rankin, J. M., 79, 80, 85, 87, 113, 114, 115, 116, 119, 123, *127, 128*
Reichley, P. E., 125, *128*
Resch, G. M., 116, *129*
Rice, S. O., 186, *219*
Richards, D. W., 114, 115, 119, 122, 123, *128*
Rickett, B. J., 78, 79, 84, 94, 96, 105, 106, 107, 109, 110, 111, 117, *127, 128, 129*
Riddle, A. C., 47, 49, 51, *53*
Roger, R. S., 172, *174*
Rogers, A. E. E., 133, *174*, 196, *209*
Ruderman, M., 125, *126*
Ryle, M., 132, 139, 147, 161, 162, 172, 173, *174, 175*

S

Salpeter, E. E., 58, 59, 112, 113, *128, 129*
Sande, G., 71, *129*
Sangster, F. L. J., 69, *129*
Scargle, J. D., 125, *129*
Schmahl, E. J., 51, *53*

Schwarz, U., 209, *218*
Scott, P. F., 56, 57, 118, *128*
Shakeshaft, J. R., 172, *173*
Sheridan, K. V., 20, 38, 46, 47, 48, 51, *53*
Shinn, D. H., 117, *127*
Shitov, Yu. P., 112, *129*
Slee, O. B., 118, *126*
Slepian, D., 168, *175*
Smerd, S. F., 4, 44, 47, 48, *53*
Smith, F. G., 84, 87, 121, *127*, *128*
Smith, S. F., 47, *53*
Smits, D. W., 209, *218*
Sondaar, L. H., 172, *173*
Squire, W. D., 72, *129*
Staelin, D. H., 58, 59, 76, 109, *127*, *129*
Stark, H., 12, *53*, 160, *174*
Stegun, I. A., 150, *173*
Steinberg, J. L., 47, *53*
Stewart, R. T., 46, 48, *53*
Stratton, J. A., 133, *175*
Sutton, J. M., 58, 59, 76, 112, *128*, *129*
Swarup, G., 14, *52*, 118, 126, 132, 172, *175*
Swenson, G. W., Jr., 133, *175*
Szebehely, V., 122, *127*

T

Tademaru, E., 77, 107, *128*
Takakubo, K., 209, *218*
Takakura, T., 14, *53*
Tanaka, H., 5, 13, 14, *53*
Tanenbaum, B. S., 80, *129*
Taylor, J. H., 57, 61, 62, 66, 76, 77, 87, 88, 89, 90, 96, 97, 102, 107, 113, 123, 124, 125, 126, *128*, *129*
Teer, K., 69, *129*
Thompson, A. R., 132, 143, 149, 164, 166, *173*, *175*
Thornton, D. D., 172, *174*
Trakhtengerts, V. Yu., 49, *53*
Tsuchiya, A., 14, *53*
Tsukiji, Y., 14, *53*
Tukey, J. W., 71, 94, 97, *127*, 164, *174*, 182, *218*

U

Uchida, Y., 48, *53*

V

Van, Y. Y., 124, 125, *128*
Vandenberg, N. R., 116, *129*
Vanderbrugge, J. F., 172, *173*
Van Someren Gréve, H. W., 162, *174*
Vanvleck, J. H., 190, *219*
Van Woerden, H., 209, *218*
Vaughan, A. E., 57, *129*
Venugopal, V. R., 118, *126*
Vinokur, M., 14, *53*
Visser, J. J., 172, *173*

W

Wade, C. M., 162, 166, 172, *174*, *175*
Wampler, J., 116, 124, *128*
Warburton, J. A., 6, 13, 52, 132, *174*
Warner, P. J., 172, *173*
Watts, D. G., 97, *128*
Weiler, K. W., 138, *175*
Weiner, N., *219*
Weiner, R., 76, *129*
Weinreb, S., 190, 191, *219*
Weiss, A. A., 46, *53*
Welch, W. J., 172, *174*
Wellington, K. J., 172, *173*
Whitehouse, H. J., 72, *129*
Whyte, D. A., 193, *218*
Wielebinski, R., 57, *129*
Wild, J. P., 12, 16, 17, 18, 19, 31, 44, 46, 47, 50, *52*, *53*
Wilkinson, D. T., 121, 125, 126, *127*
Williamson, I. P., 113, *129*
Wilson, D. M. A., 172, *173*
Wolf, E., 133, *173*, 212, *218*
Wright, M. C. H., 172, *173*

Y

Yao, S. S., 168, *174*
Yen, J. L., 196, *218*
Youmans, A. B., 116, *129*

Z

Zeissig, G. A., 80, *129*
Zheleznyakov, V. V., 49, *53*

Subject Index

A

Aberration, 144
Aliasing, 82, 99, 165
Amplitude modulated noise, 91
  model, 107
Angular resolution, 13, 49
Antenna
  linearly polarized, 120
  log-periodic dipole, 16
  pattern, 5, 149
    of paraboloid, 154
Aperture synthesis, 13, 132
  telescope, 159, 172
Arecibo Observatory, 60
Array
  circular, 10, 15
  cross, 10, 159
  fan-beam, 13
  five Km, 132
  grating, 13
  low redundancy, 10
  matrix, 8
  moving element, 141
  n-element, 18
  orthogonally polarized, 56, 85
  polarization response, 81
  ring- shaped, 132
  spaced, 6
  steering, 22
  swept-frequency grating, 14
  "T", 14
  very long, 173
Arrival-time curvature, 67
Autocorrelation, 10, 111, 117, 215
  function, 134, 136, 143, 178, 193
  hard clipped, 190
  one-bit, 190, 204
  spectrometer, 169, 189, 190, 204
Autocovariance, 94
  function, 91
Automatic gain control, 23, 207
  filter spectrometer, 206
Autoregression, 169

B

Bandpass, 68, 111, 192, 202
  filter, 180
  noise spectrum, 204
Bandwidth, 11, 67, 91, 107, 109, 116, 152, 163, 189, 197
Barycentric arrival times, 122
Baseline, 149
  curvature, 205
  error, 162
Beam efficiency, 200
Beam pattern, 147, 153
  Gaussian, 201
  pencil, 6, 18, 21, 26
  primary, 154, 165
Beam steering, 21
Beamwidth, 87
  half power, 148, 159
Bessel function, 16, 148
Binary phase switches, 24
Binary tree algorithm, 59, 63
Bragg cell spectrometer, 188, 189, 197
Braking index, 125
Branching network, 13, 21
Brightness, 134, 136, 138, 144, 159
  apparent, 29
  distribution, 109, 137, 163
    Gaussian, 34
  function, 169
  map, 166
  scale, 149
  sensitivity, 159
  temperature, 156, 200, 204
  threshold, 33
Bucket-brigade principle, 69
Bursts, 46

C

Calibration procedure, 202
Calibration system, 162
Cathode ray tube, 23, 197
Celestial sphere, 140

Cell summing, 165
Central lobe, 16
Centroid of source position, 36, 78
Charge-coupled devices, 72
Chebyshev polynomials, 210
Chi-squared distribution, 103
  statistics, 93, 98
Chromosphere, 2
Circularly polarized, 45, 120
"Cleaning" process, 170
  convergence, 171
Collimation error, 149
Complex envelope function, 82
Complex multiplication and addition, 163
Computer editing, 29
Computing time, 166
Concentric antennas, 84
Confusion, 158
Continuous filled-aperture array, 7
Continuum radiation, 142
Contour diagrams, 170
Control Computer, 23
Convolution, 8, 68, 72, 134, 137, 167, 180, 184, 195, 215
  inversion, 215
Coordinate shift, 152
Corona, 3, 5, 16, 46
Correlation, enhanced, 96
  one-bit, 192, 206
Correlation functions, 8, 93, 96, 137, 139, 141, 152, 156, 160, 161, 163, 165, 170
  length, 147
Correlator
  multibit, 192
Cosine transform, 179, 192
Crab Nebula, 115
Crab pulsar, see Pulsar, PSR 0531 + 21
Cross-correlation, 11, 14, 61, 105, 106, 112, 196
  coefficients, 117
  function, 78, 119

D

Data compression, 40
Deconvolution, 112
Delay line, 23, 69
Delays, 69
Dielectric constant, 134

Difference pattern, 157
Diffraction angle, 189
  grating, 5
  pattern, 16, 131
Digital dispersion, 62
  filtering, 71
Dirac delta function, 137, 165
Directivity, 138
Dish sensitivity, 159
Dispersion coefficient, 67, 77, 78, 80, 120
  delay, 61, 114
  distortion, 64, 70
  interstellar, 57, 64
  removal, 82
Doppler frequency distribution, 217
  shift, 79, 83, 121
  velocity, 202, 207, 214
Drift velocity, 117
Drifting spectral bands, 102, 109
Duty factor, 64
Dynamic range of system, 23, 157, 173
Dynamic spectra, 44, 197

E

Earth atmosphere, 161
  elasticity, 144
  diurnal aberration, 144
  nutation, 144
  precession, 144
    variation, 161
  rotation, 139, 144, 208
    nonuniform, 161
    parameters, 162
  seismic motion, 195
  spin, 83
  tides, 146
  orbit, 122, 139
  velocity, 79
Eigenvalue, 168
Electric field, 134, 136, 139
Electromagnetic field, 133
Electron density, 2, 64, 81
Electrons, 45, 46, 49, 66, 80, 109
Elongated sources, 44
Energy flux, 106
Entropy, 169
  maximum, 193
Ergodicity, 135
Error pattern, 152

F

Fan-beam, 49, 132
  mode, 10
Faraday rotation, 81, 86
  differential, 85
Fast folding algorithm, 58
Fast Fourier transform, 58, 60, 164, 192
  machine, 196
Filter, finite width, 179
  rail, 58, 61
  scanning, 179, 184
  software, 75
  transversal, 72
  variable frequency, 184
Flare explosion, 47
Fluctuations
  intrinsic, 102
  periodic, 97
  random, 93
  signal, 64, 69
Flux, 155
  density, 32, 201
  rms error, 154
FORTRAN, 209
Fourier coefficients, 80, 114
  components, 12, 18
  integral, 179
  theory, 133
  transform, 7, 8, 65, 67, 72, 97, 108, 112, 136, 148, 168, 178, 190, 215
    inverse, 193
    three-dimensional, 141, 144, 163
    two-dimensional, 141, 144, 163
Fourier transform, see also Fast Fourier transform
  direct, 196
  inverse, 193
Fractional gain error, 203, 204
Fraunhofer diffraction, 12
Frequency offsets, 207
Frequency-programming unit, 26
Frequency resolution, 106
  sweep rate, 66
Fresnel pattern, 68
Fringe stopping, 140, 145, 161
  center, 148, 152

G

Gain errors, 203, 204
Galactic plane, 57

Gaussian noise, 58, 69, 154
Gaussian random variable, 190
Gaussian-shaped beam, 171
Glitch function, 126
Grading function, 163
Grating response, 16
  rings, 143, 149, 155
  sidelobes, 18, 170
Grid, optimum, 166
Gridding procedures, 165

H

Halftone display, 30
Hanning weighting, 192
Hatbox function, 153
Humidity, 147
Hyperfine structure, 214, 217

I

Impulse response, 68, 82
Incoherent addition of intensities, 213
  radiation, 135
Instrumental corrections, 79
Integral, 142
Integration, 18, 31, 42, 96, 114, 116, 121, 136, 156
  interval, 119
  off-pulse, 94
  quiet-Sun, 44
Integrator (averager), 188
Interferometer, 116, 146
  compound, 13
Interference pattern, 10
Intermediate frequency, 76, 142
  amplifier, 79
  channel, 29
Interplanetary scintillation, 117
Interpolation, 35
Interstellar environment, 178
  scattering, 64, 82, 114
  scintillation, 57, 104, 109, 112, 120
  velocity, 117
Intrapulse time, 87
Instrinsic profile, 114
Instrumental errors, 162
Ionized gas, 3
Ionosonde profiles, 86
Ionosphere, 3, 39, 146
  refraction, 147
  scintillation, 117

J

"J^2-synthesis,", 10, 18, 24
Jodrell Bank, 83

L

Lag, 179, 193
Lagged products, 93
Lagrangian multipliers, 194
Laguerre functions, 115
Least overall aliasing, 167
Least-squared error fitting, 114, 123
   Chebyshev, 209
   Gaussian, 208
   polynomial, 208
Lightning flashes, 43
Line shape, 217
Linear aperture, 5
Linear interpolation, 30
Local oscillator, 75, 76
Long period features, 98
Longitudinal filters, 74
Lunar occulation, 68

M

Magnetic core store, 26
Magnetic dipole braking, 123
Magnetic fields, 4, 44, 45, 46, 80, 81, 84, 86
Magnetic tapes, 28, 38, 40, 64, 71, 161
Magnetohydrodynamic waves, 2
Magnetosphere, 88, 107
Matrix inversion, 209
Mechanical phase shifters, 24
Mode changes, 121
Molecular spectrum, 217
Moon, 144
Multiplication, 184
   one-bit, 190
Multiscaler, 83

N

Nail-bed function, 165
National Radio Astronomy Observatory, 166
Neutron stars, 84, 93, 125, 126

Noise, 154, 156, 206
   calibration source, 204
   intensity, 94
   receiver, 107
   temperature, 89, 155
   uncorrelated, 155
   white Gaussian, 91, 179, 186
Nonequilibrium populations of levels, 178
Notch, 77, 101
Nulling phenomenon, 98
Nyquist criterion, 192
   frequency, 82, 99
   limit, 61
   sampling theorem, 180, 185

O

Opacity, 214
Optical processor, 11
Optoacoustical processor, 72

P

Parabolic reflector, 16
PDP-15, 42
Pencil-beam radioheliograph, 14
   resolution, 15
Period resolution, 58
Phase control, 147
   system, 16
   correlation, 113
   error, 142, 149
   lag, 146
   noise, 154
   relative, 78
   screen model, 112
   shifts, 14, 18, 21, 66, 137
   steering, 14
   sweeping, 14
   switching, 84
Photodiode array, 189
Photographic display, 199
Photon-phonon collisions, 189
Photosphere, 2, 16
Picture-point, 11, 12, 20, 21, 23, 26
Piezoelectric transducer, 188
Planck blackbody law, 201
Plane wave, 11
Plasma, 44, 47, 48, 64, 65, 80, 115
Poincaré sphere, 213

SUBJECT INDEX 229

Polar diagram, 10, 18, 26
Polar motion, 145, 161
Polarization, 22, 32, 40, 51, 75, 84, 106, 120, 138, 212
  circular
    left-handed, 37
    right-handed, 36
  elliptical, 87, 214
  fractional, 213
  orthogonal, 81
  pulsar, 84
Polarization errors, 120
  fluctuation, 87
  parameters, 213
  profile, 86
Position angle correction, 86
Postdetection dispersion removal, 76
  filter, 81, 114
  smoothing, 68
  time constant, 82
Postdetector response, 91
Power pattern, 155
  polar diagram, 6
  angular spectrum, 112
  response, 7
  spectrum, 75, 97, 107, 112, 178, 184, 193
    coefficients, 61
  transfer function, 179
Poynting vector, 134
Predetection dispersion removal, 84
Pressure, 147
Probability distribution function, 102
Prolate spheroidal function, 168
Pulsar, 87, 177, 197, 199
  CPO 329, 199
  PSR 0031-07, 102
  PSR 0329+54, 110, 117, 121
  PSR 0531+21, 76, 79, 80, 104, 113, 116, 124
  PSR 0531+22, 114
  PSR 0809+74, 91, 102
  PSR 0833-45, 86, 118, 125
  PSR 0834+06, 102
  PSR 0950+08, 69, 77, 78, 91, 105
  PSR 1133+16, 77, 124
  PSR 1237+25, 91, 100, 121
  PSR 1237+28, 99
  PSR 1508+55, 126
  PSR 1642−03, 91
  PSR 1919+21, 57, 77, 78, 98, 101, 102
  PSR 2016+28, 94, 97, 102, 105
  PSR 2303+30, 102
Pulsar intensity, 68
  micropulses, 77
  microstructure, 71
  period, 83
  polarization, 84
  rotating, 87
  scintillation, 117
  signals, 56, 64
Pulse averaging, 56
  average intrinsic shape, 112
  average profile, 56, 57, 88
  broadening, 116
  depolarization, 86
  energy, 104
  fluctuations, 76, 97
  height analyzer, 83
  intensity, 112
  period ambiguity, 78
  polarization, 86
  profile, 78, 91, 99, 103, 121
    error, 103
  rate converter, 25
  spectrum, 111
Pulse-to-pulse correlation, 103
  variability, 118

Q

Quiet Sun, 31, 40

R

Radial coordinates, 148
Radio brightness, see Brightness
Radio interference, 42
  noise, 131
  pulse intensity, 56
  radiation, 157
  spectrograph, 137
  source, 136, 155
    celestial, 161
    extended, 171, 201
Radio source, 136, 155, 171, 201
  burst, 40
  celestial, 161
  extended, 171, 201
  individual, 158
  point, 140, 157, 201

small diameter, 201
quasipoint, 171
scintillating, 56
storm, 37
weak, 158
Random variable, 112
Raster, 40, 42
Ray path, 146
Rayleigh-Jeans approximation, 201
Real time, 40, 72, 75
search, 64
Rectangular pulse train, 61
Refraction, 146
differential, 163
Relativistic correction, 122
Resolution filter, 186
Resolved spectrum, 195
Resolving power, 131, 132, 158, 169
Rotation synthesis, 132, 139, 141

S

Sampling function, two dimensional, 7
interval, 153
Shift theorem, 66
Shock front, 49
waves, 47
Sidelobes, 8, 18, 33, 72, 147, 148, 152, 156, 157, 165, 169, 170, 181
aliased, 170
negative, 155
ratio, 33
spectral, 195
Signal enhancement, 69
Signal-to-noise ratio, 31, 61, 69, 76, 82, 89, 104, 168, 195
Single pulse search technique, 62
Skirts, 41
Smearing function, 83
Smoothing, 81, 98, 156, 181
Solar map, two dimensional, 14
Spectral estimate, 194
Spectral power, 99
Spectrometer
autocorrelation, 169, 189, 190, 204
multichannel, 76, 187
scanning filter, 186, 197
Square-law detection, 66
Square-law detector, 178
power, 184

Solar bursts, 13, 28, 31, 35, 44, 197 (see also Bursts)
corona, 80
disk, 37
emission, 1
storm, 35
Spectral-line radio astronomy, 195
narrow, 184
Spectrum coefficients, 61
Split-frequency spectrum, 48
Square aperture, 6
Square filter, 179, 185
Square-law detector, 68
power, 188
Staircase function, 22
Standard deviation, 93
Stationary phase, 68
integral, 75
Stokes parameters, 84, 86, 212
Stripscans, 169
Subpulse bands, 101
Sun, 144
Super-synthesis, 139
Surface wave delay lines, 72
Sweep rate, 75
Switching times, 84
Synchrotron radiation, 49, 51
Synthesis map, 149
telescope, 158, 160
Synthesized beam, 159
pattern, 168
System temperature, 202, 204

T

Tapering function, 143, 166
Taylor series, 65, 123, 208
expansion, 210
Telescope
line array synthesis, 164
rotational aperture synthesis, 141
single element, 161
Temperature, 146
Test dipole, 86
Thermal background emission, 44
Thermonuclear burning, 2
Thunderstorms, 43
Time averaging, 84
Time constant function, 153
Time delay, 66, 78, 117, 142, 152

Time resolution, 14, 57, 67, 71, 76, 82, 106, 119
Toeplitz matrix, 195
Tracking counter, 24, 25
Transfer function, 65, 71, 72
Two-dimensional acv, 96
Two-screen model, 115

V

Variable time scale, 30
Variations of pulse intensity, 87

Vela pulsar, see Pulsar PSR 0833—45
Very long baseline interferometry, 116, 132
Vidicon tube, 189
Visibility function, 116
Voltage polar diagrams, 8

W

Weighting function, 155, 163, 180, 182
  uniform, 195

Contents of Previous Volumes

Volume 1: Statistical Physics

The Numerical Theory of Neutron Transport
Bengt G. Carlson

The Calculation of Nonlinear Radiation Transport by a Monte Carlo Method
Joseph A. Fleck, Jr.

Critical-Size Calculations for Neutron Systems by the Monte Carlo Method
Donald H. Davis

A Monte Carlo Calculation of the Response of Gamma-Ray Scintillation Counters
Clayton D. Zerby

Monte Carlo Calculation of the Penetration and Diffusion of Fast Charged Particles
Martin J. Berger

Monte Carlo Methods Applied to Configurations of Flexible Polymer Molecules
Frederick T. Wall, Stanley Windwer, and Paul J. Gans

Monte Carlo Computations on the Ising Lattice
L. D. Fosdick

A Monte Carlo Solution of Percolation in the Cubic Crystal
J. M. Hammersley

AUTHOR INDEX—SUBJECT INDEX

Volume 2: Quantum Mechanics

The Gaussian Function in Calculations of Statistical Mechanics and Quantum Mechanics
Isaiah Shavitt

Atomic Self-Consistent Field Calculations by the Expansion Method
C. C. J. Roothaan and P. S. Bagus

The Evaluation of Molecular Integrals by the Zeta-Function Expansion
M. P. Barnett

Integrals for Diatomic Molecular Calculations
Fernando J. Corbató and Alfred C. Switendick

Nonseparable Theory of Electron-Hydrogen Scattering
A. Temkin and D. E. Hoover

Estimating Convergence Rates of Variational Calculations
Charles Schwartz

Author Index—Subject Index

Volume 3: Fundamental Methods in Hydrodynamics

Two-Dimensional Lagrangian Hydrodynamic Difference Equations
William D. Schulz

Mixed Eulerian-Lagrangian Method
R. M. Frank and R. B. Lazarus

The Strip Code and the Jetting of Gas between Plates
John G. Trulio

CEL: A Time-Dependent, Two-Space-Dimensional, Coupled Eulerian-Lagrange Code
W. F. Noh

The Tensor Code
G. Maenchen and S. Sack

Calculation of Elastic-Plastic Flow
Mark L. Wilkins

Solution by Characteristics of the Equations of One-Dimensional Unsteady Flow
N. E. Hoskin

The Solution of Two-Dimensional Hydrodynamic Equations by the Method of Characteristics
D. J. Richardson

The Particle-in-Cell Computing Method for Fluid Dynamics
Francis H. Harlow

The Time-Dependent Flow of an Incompressible Viscous Fluid
Jacob Fromm

AUTHOR INDEX—SUBJECT INDEX

Volume 4: Applications in Hydrodynamics

Numerical Simulation of the Earth's Atmosphere
Cecil E. Leith

Nonlinear Effects in the Theory of a Wind-Driven Ocean Circulation
Kirk Bryan

Analytic Continuation Using Numerical Methods
Glenn E. Lewis

Numerical Solution of the Complete Krook-Boltzmann Equation for Strong Shock Waves
Moustafa T. Chahine

The Solution of Two Molecular Flow Problems by the Monte Carlo Method
J. K. Haviland

Computer Experiments for Molecular Dynamics Problems
R. A. Gentry, F. H. Harlow, and R. E. Martin

Computation of the Stability of the Laminar Compressible Boundary Layer
Leslie M. Mack

Some Computational Aspects of Propeller Design
William B. Morgan and John W. Wrench, Jr.

Methods of the Automatic Computation of Stellar Evolution
Louis G. Henyey and Richard D. Levée

Computations Pertaining to the Problem of Propagation of a Seismic Pulse in a Layered Solid
F. Abramovici and Z. Alterman

AUTHOR INDEX—SUBJECT INDEX

Volume 5: Nuclear Particle Kinematics

Automatic Retrieval Spark Chambers
J. Bounin, R. H. Miller, and M. J. Neumann

Computer-Based Data Analysis Systems
Robert Clark and W. F. Miller

Programming for the PEPR System
P. L. Bastien, T. L. Watts, R. K. Yamamoto, M. Alston, A. H. Rosenfeld, F. T. Solmitz, and H. D. Taft

A System for the Analysis of Bubble Chamber Film Based upon the Scanning and Measuring Projector (SMP)
Robert I. Hulsizer, John H. Munson, and James N. Snyder

A Software Approach to the Automatic Scanning of Digitized Bubble Chamber Photographs
Robert B. Marr and George Rabinowitz

AUTHOR INDEX—SUBJECT INDEX

Volume 6: Nuclear Physics

Nuclear Optical Model Calculations
Michael A. Melkanoff, Tatsuro Sawada, and Jacques Raynal

Numerical Methods for the Many-Body Theory of Finite Nuclei
Kleber S. Masterson, Jr.

Application of the Matrix Hartree-Fock Method to Problems in Nuclear Structure
R. K. Nesbet

Variational Calculations in Few-Body Problems with Monte Carlo Method
R. C. Herndon and Y. C. Tang

Automated Nuclear Shell-Model Calculations
S. Cohen, R. D. Lawson, M. H. Macfarlane, and M. Soga

Nucleon-Nucleon Phase Shift Analyses by Chi-Squared Minimization
Richard A. Arndt and Malcolm H. MacGregor

AUTHOR INDEX—SUBJECT INDEX

Volume 7: Astrophysics

The Calculation of Model Stellar Atmospheres
Dimitri Mihalas

Computational Methods for Non-LTE Line-Transfer Problems
D. G. Hummer and G. Rybicki

Methods for Calculating Stellar Evolution
R. Kippenhahn, A. Weigert, and Emmi Hofmeister

Computational Methods in Stellar Pulsation
R. F. Christy

Stellar Dynamics and Gravitational Collapse
Michael M. May and Richard H. White

AUTHOR INDEX—SUBJECT INDEX

Volume 8: Energy Bands of Solids

Energy Bands and the Theory of Solids
J. C. Slater

Interpolation Schemes and Model Hamiltonians in Band Theory
J. C. Phillips and R. Sandrock

The Pseudopotential Method and the Single-Particle Electronic Excitation Spectra of Crystals
David Brust

A Procedure for Calculating Electronic Energy Bands Using Symmetrized Augmented Plane Waves
L. F. Mattheiss, J. H. Wood, and A. C. Switendick

Interpolation Scheme for the Band Structure of Transition Metals with Ferromagnetic and Spin-Orbit Interactions
Henry Ehrenreich and Laurent Hodges

Electronic Structure of Tetrahedrally Bonded Semiconductors: Empirically Adjusted OPW Energy Band Calculations
Frank Herman, Richard L. Kortum, Charles D. Kuglin, John P. Van Dyke, and Sherwood Skillman

The Green's Function Method of Korringa, Kohn, and Rostoker for the Calculation of the Electronic Band Structure of Solids
Benjamin Segall and Frank S. Ham

AUTHOR INDEX—SUBJECT INDEX

Volume 9: Plasma Physics

The Electrostatic Sheet Model for a Plasma and Its Modification to Finite-Sized Particles
John M. Dawson

Solution of Vlasov's Equation by Transform Methods
Thomas P. Armstrong, Rollin C. Harding, Georg Knorr, and David Montgomery

The Water-Bag Model
Herbert L. Berk and Keith V. Roberts

The Potential Calculation and Some Applications
R. W. Hockney

Multidimensional Plasma Simulation by the Particle-in-Cell Method
R. L. Morse

Finite-Size Particle Physics Applied to Plasma Simulation
Charles K. Birdsall, A. Bruce Langdon, and H. Okuda

Finite-Difference Methods for Collisionless Plasma Models
Jack A. Byers and John Killeen

Application of Hamilton's Principle to the Numerical Analysis of Vlasov Plasmas
H. Ralph Lewis

Magnetohydrodynamic Calculations
Keith V. Roberts and D. E. Potter

The Solution of the Fokker–Planck Equation for a Mirror-Confined Plasma
John Killeen and Kenneth D. Marx

AUTHOR INDEX—SUBJECT INDEX

Volume 10: Atomic and Molecular Scattering

Numerical Solutions of the Integro-Differential Equations of Electron–Atom Collision Theory
P. G. Burke and M. J. Seaton

Quantum Scattering Using Piecewise Analytic Solutions
Roy G. Gordon

Quantum Calculations in Chemically Reactive Systems
John C. Light

Expansion Methods for Electron–Atom Scattering
Frank E. Harris and H. H. Michels

Calculation of Cross Sections for Rotational Excitation of Diatomic Molecules by Heavy Particle Impact: Solution of the Close-Coupled Equations
William A. Lester, Jr.

Amplitude Densities in Molecular Scattering
Don Secrest

Classical Trajectory Methods
Don L. Bunker

AUTHOR INDEX—SUBJECT INDEX

Volume 11: Seismology: Surface Waves and Earth Oscillations

Finite Difference Methods for Seismic Wave Propagation in Heterogeneous Materials
David M. Boore

Numerical Analysis of Dispersed Seismic Waves
A. M. Dziewonski and A. L. Hales

Fast Surface Wave and Free Mode Computations
F. A. Schwab and L. Knopoff

A Finite Element Method for Seismology
John Lysmer and Lawrence A. Drake

Seismic Surface Waves
H. Takeuchi and M. Saito

AUTHOR INDEX—SUBJECT INDEX

Volume 12: Seismology: Body Waves and Sources

Numerical Methods of Ray Generation in Multilayered Media
F. Hron

Computer Generated Seismograms
Z. Alterman and D. Loewenthal

Diffracted Seismic Signals and Their Numerical Solution
C. H. Chapman and R. A. Phinney

Inversion and Inference for Teleseismic Ray Data
Leonard E. Johnson and Freeman Gilbert

Multipolar Analysis of the Mechanisms of Deep-Focus Earthquakes
M. J. Randall

Computation of Models of Elastic Dislocations in the Earth
Ari Ben-Menahem and Sarva Jit Singh

AUTHOR INDEX—SUBJECT INDEX

Volume 13: Geophysics

Signal Processing and Frequency-Wavenumber Spectrum Analysis for a Large Aperture Seismic Array
Jack Capon

Models of the Sources of the Earth's Magnetic Field
Charles O. Stearns and Leroy R. Alldredge

Computations with Spherical Harmonics and Fourier Series in Geomagnetism
D. E. Winch and R. W. James

Inverse Methods in the Interpretation of Magnetic and Gravity Anomalies
M. H. P. Bott

Analysis of Geoelectromagnetic Data
S. H. Ward, W. J. Peeples, and J. Ryu

Nonlinear Spherical Harmonic Analysis of Paleomagnetic Data
J. M. Wells

Harmonic Analysis of Earth Tides
Paul Melchior

Computer Usage in the Computation of Gravity Anomalies
Manik Talwani

Analysis of Irregularities in the Earth's Rotation
D. E. Smylie, G. K. C. Clarke, and T. J. Ulrych

Convection in the Earth's Mantle
D. L. Turcotte, K. E. Torrance, and A. T. Hsui

AUTHOR INDEX—SUBJECT INDEX